LIGHT–MATTER INTERACTION

LIGHT—MATTER INTERACTION

Volume 1

Fundamentals and Applications

John Weiner
Laboratoire de Collisions, Agrégats et Réactivité
Université Paul Sabatier

P.-T. Ho
Department of Electrical and Computer Engineering
University of Maryland

A JOHN WILEY & SONS PUBLICATION

Published by John Wiley & Sons, Inc., Hoboken, New Jersey.
Published simultaneously in Canada.

For general information on our other products and services please contact our Customer Care Department within the U.S. at 877-762-2974, outside the U.S. at 317-572-3993 or fax 317-572-4002.

Wiley also publishes its books in a variety of electronic formats. Some content that appears in print, however, may not be available in electronic format.

Library of Congress Cataloging-in-Publication Data

Weiner, John, 1943–
Light–matter interaction / John Weiner, Ping-Tong Ho.
p. cm.
Includes bibliographical references and index.
Contents: v. 1. Fundamental and applications
ISBN 0-471-25377-4 (v. 1 : acid-free paper)
1. Atoms. 2. Physics. 3. Molecules. 4. Optics. I. Ho, Ping-Tong. II. Title.

QC173 .W4325 2003
539—dc21

Printed in the United States of America.

10 9 8 7 6 5 4 3 2 1

CONTENTS

PREFACE

Atomic, molecular, and optical (AMO) science and engineering is at the intersection of strong intellectual currents in physics, chemistry, and electrical engineering. It is identified by the research community responsible for fundamental advances in our ability to use light to observe and manipulate matter at the atomic scale, use nanostructures to manipulate light at the subwavelength scale, develop new quantum-electronic devices, control internal molecular motion and modify chemical reactivity with pulsed light.

This book is an attempt to draw together principal ideas needed for the practice of these disciplines into a convenient treatment accessible to advanced undergraduates, graduate students, or researchers who have been trained in one of the conventional curricula of physics, chemistry, or engineering but need to acquire familiarity with adjacent areas in order to pursue their research goals.

In deciding what to include in the volume we have been guided by a simple question: "What was missing from our own formal education in chemical physics or electrical engineering that was indispensable for a proper understanding of our AMO research interests"? The answer was: "Plenty!", so this question was a necessary but hardly sufficient criterion for identifying appropriate material.

The choices therefore, while not arbitrary, are somewhat dependent on our own personal (sometimes painful) experiences. In order to introduce essential ideas without too much complication we have restricted the treatment of microscopic light-matter interaction to a two-level atom interacting with a single radiation field mode. When a gain medium is introduced, we treat real lasers of practical importance. While the gain medium is modeled as three- or four-level systems, it can be simplified to a two-level system in calculating the important physical quantities. Wave optics is treated in two dimensions in order to prevent elaborate mathematical expressions from obscuring the basic physical phenomena. Extension to three dimensions is usually straightforward; and when it is, the corresponding results are given.

Chapter 1 introduces the consequences of an ensemble of classical, radiating harmonic oscillators in thermal equilibrium as a model of blackbody radiation and the phenomenological Einstein rate equations with the celebrated A and B coefficients for the absorption and emission of radiation by matter. Although the topics treated are "old fashioned" they set the stage for the quantized oscillator treatment of the radiation field in Chapter 5 and the calculation of the B coefficient from a simple semiclassical model in Chapter 2. We have found in teaching this material that students are seldom acquainted with density matrices, essential for the treatment of the optical Bloch equations (OBEs). Therefore chapter 3 outlines the essential properties of density matrices before discussing the OBEs applied to a two-level atom in Chapter 4. We treat light-matter interaction macroscopically in terms of dielectric polarization and susceptibility in Chapter 4 and show that, aside from spontaneous emission, light-matter energies and forces need not be considered intrinsically quantal. Energies and forces are derived from the basic Lorentz driven-oscillator model of the atom interacting with a classical optical field. This picture is more "tangible" than the formalism of quantum mechanics and helps students get an intuitive grasp of much, if not all, light-matter phenomena. In Chapter 7 and its appendixes we develop this picture more fully and point out analogies to electrical circuit theory. This approach is already familiar to students with an engineering background but perhaps less so to physicists and chemists. Chapter 5 does quantize the field and then develops "dressed states" which put atom or molecule quantum states and photon number states on an equal footing. The dressed-state picture of atom-light interaction is a time-independent approach that complements the usual time-dependent driven-oscillator picture of atomic transitions and forces. Chapters 6 and 7 apply the tools developed in the preceding chapters to optical methods of atom trapping and cooling and to the theory of the laser. Chapter 8 presents the fundamentals of geometric and wave optics with applications to typical laboratory situations. Chapters 6, 7, and 8 are grouped together as "Applications" because these chapters are meant to bring theory into the laboratory and show students that they can use it to design and execute real experiments. The only way to really master this material and make it useful to the reader is to work out applications to realistic laboratory situations. Furthermore, sometimes the easiest and clearest way to present new material is by examples. For these reasons we have seeded the text with quite a few *Problems* and *Examples* to complement the formal presentation.

Special acknowledgment is due to Professor William DeGraffenreid, for his skill and patience in executing all the figures in this book. It has been a pleasure to have him first as a student then as a colleague since 1997. Thanks are also due to students too numerous to mention individually who in the course of teacher–student interaction at the University of Maryland and at l'Université Paul Sabatier, Toulouse revealed and corrected many errors in this presentation of light–matter interaction.

We have tried to organize key ideas from the relevant areas of AMO physics and engineering into a format useful to students from diverse backgrounds working in an inherently multidisciplinary area. We hope the result will prove useful to readers and welcome comments, and suggestions for improvement.

John Wiener
Toulouse, France

P.-T. Ho
College Park, Maryland

Part I

Light-Matter Interaction: Fundamentals

Chapter 1

Absorption and Emission of Radiation

1.1 Radiation in a Conducting Cavity

1.1.1 Introduction

In the age of lasers it might be legitimately asked why it is still worthwhile to bother with classical treatments of the emission and absorption of radiation. There are several reasons. First, it deepens our physical understanding to identify exactly how and where a perfectly sound classical development leads to preposterous results. Second, even with narrowband, monomode, phase-coherent radiation sources, the most physically useful picture is often a classical optical field interacting with a quantum-mechanical atom or molecule. Third, the treatment of an ensemble of classical oscillators subject to simple boundary conditions prepares the analogous development of an ensemble of quantum oscillators and provides the most direct and natural route to the quantization of the radiation field.

Although we seldom perform experiments by shining light into a small hole in a metal box, the field solutions of Maxwell's equations are particularly simple for boundary conditions in which the fields vanish at the inner surface of a closed structure. Before discussing the physics of radiation in such a perfectly conducting cavity, we introduce some key relations between electromagnetic field amplitudes, the stored field energy, and the intensity. A working familiarity with these relations will help us develop important results that tie experimentally measurable quantities to theoretically meaningful expressions.

1.1.2 Relations among classical field quantities

Since virtually all students now learn electricity and magnetism with the rationalized mks system of units, we adopt that system here. This choice means that we write Coulomb's force law between two electric charges q, q' separated by a distance r as

$$\mathbf{F} = \frac{1}{4\pi\epsilon_0}\left(\frac{qq'}{r^3}\right)\mathbf{r} \tag{1.1}$$

and Ampère's force law (force per unit length) of magnetic induction between two infinitely long wires carrying electric currents I, I', separated by a distance r as

$$\frac{d\,|\mathbf{F}|}{d\,l} = \frac{\mu_0}{2\pi}\left(\frac{II'}{r}\right) \tag{1.2}$$

where ϵ_0 and μ_0 are called the *permittivity of free space* and the *permeability of free space*, respectively. In this units system the permeability of free space is defined as

$$\frac{\mu_0}{4\pi} \equiv 10^{-7} \tag{1.3}$$

and the numerical value of the permittivity of free space is fixed by the condition that

$$\frac{1}{\epsilon_0\mu_0} = c^2 \tag{1.4}$$

Therefore we must have

$$\frac{1}{4\pi\epsilon_0} = 10^{-7}c^2 \tag{1.5}$$

The electric field of the standing-wave modes within a conducting cavity in vacuum can be written

$$\mathbf{E} = \mathbf{E_0}e^{-i\omega t}$$

where $\mathbf{E_0}$ is a field with amplitude E_0 and a polarization direction $\mathbf{e}$. The $\mathbf{E_0}$ field is transverse to the direction of propagation, and the polarization vector resolves into two orthogonal components. The magnetic induction field amplitude associated with the wave is B_0, and the relative amplitude between magnetic and electric fields is given by

$$B_0 = \frac{1}{c}E_0 = \sqrt{\epsilon_0\mu_0}E_0 \tag{1.6}$$

The quantity k is the amplitude of the wave vector and is given by

$$k = \frac{2\pi}{\lambda}$$

with λ the wavelength and ω the angular frequency of the wave. For a traveling wave the $\mathbf{E}$ and $\mathbf{B}$ fields are in phase but as a standing wave they are out of phase.

The energy of a standing-wave electromagnetic field, oscillating at frequency ω, and *averaged over a cycle of oscillation*, is given by

$$\bar{U}_\omega = \frac{1}{2}\int\frac{1}{2}\left(\epsilon_0\,|\mathbf{E}|^2 + \frac{1}{\mu_0}\,|\mathbf{B}|^2\right)dV$$

and the spectral energy *density*, by

$$\frac{d\bar{U}_\omega}{dV} = \bar{\rho}_\omega = \frac{1}{4}\left(\epsilon_0 \left|\mathbf{E}\right|^2 + \frac{1}{\mu_0}\left|\mathbf{B}\right|^2\right)$$

From Eq. 1.6 we see that the electric field and magnetic field contribution to the energy are equal. Therefore

$$\bar{U}_\omega = \frac{1}{2}\int \epsilon_0 \left|\mathbf{E}\right|^2 dV \tag{1.7}$$

and

$$\bar{\rho}_\omega = \frac{1}{2}\epsilon_0 \left|\mathbf{E}\right|^2 \tag{1.8}$$

When considering the standing-wave modes of a cavity, we are interested in the spectral energy density $\bar{\rho}_\omega$, but when considering traveling-wave light sources such as lamps or lasers, we need to take account of the spectral width of the source. We define the energy density $\bar{\rho}$ as the spectral energy density $\bar{\rho}_\omega$ integrated over the spectral width of the source

$$\bar{\rho} = \int_{\omega_0 - \frac{\Delta\omega}{2}}^{\omega_0 + \frac{\Delta\omega}{2}} \bar{\rho}_\omega d\omega = \int_{\omega_0 - \frac{\Delta\omega}{2}}^{\omega_0 + \frac{\Delta\omega}{2}} \frac{d\bar{\rho}}{d\omega} d\omega$$

so

$$\bar{\rho}_\omega = \frac{d\bar{\rho}}{d\omega} \tag{1.9}$$

Another important quantity is the flow of electromagnetic energy across a boundary. The Poynting vector describes this flow, and is defined in terms of $\mathbf{E}$ and $\mathbf{B}$ by

$$\mathbf{I} = \frac{1}{\mu_0}\left(\mathbf{E} \times \mathbf{B}\right)$$

Again taking into account Eq. 1.6, we see that the magnitude of the period-averaged Poynting vector is

$$\bar{I} = \frac{1}{2}\epsilon_0 c \left|\mathbf{E}\right|^2 \tag{1.10}$$

The magnitude of the Poynting vector is usually called the *intensity* of the light, and it is consistent with the idea of a flux being equal to a density multiplied by a speed of propagation. Just as for the field energy density, we distinguish a spectral energy flux $\bar{I}_\omega$ from the energy flux $\bar{I}$ integrated over the spectral width of the light source:

$$\bar{I}_\omega = \frac{d\bar{I}}{d\omega}$$

From Eq. 1.8 for the spectral energy density of the field we see that in the direction of propagation with velocity c the spectral energy flux in vacuum would be

$$\Im = \bar{\rho}_\omega c = \frac{1}{2}\epsilon_0 c \left|\mathbf{E}\right|^2 = \bar{I}_\omega \tag{1.11}$$

which is the same expression as the magnitude of the period-averaged Poynting vector in Eq. 1.10. The spectral intensity can also be written as

$$\bar{I}_\omega = \frac{1}{2}\sqrt{\frac{\epsilon_0}{\mu_0}}\,|\mathbf{E}|^2 \tag{1.12}$$

where the factor

$$\sqrt{\frac{\mu_0}{\epsilon_0}}$$

is sometimes termed "the impedance of free space" R_0 because it has units of resistance and is numerically equal to 376.7 Ω , a factor quite useful for practical calculations. Equation 1.12 bears an analogy to the power dissipated in a resistor

$$W = \frac{1}{2}\frac{V^2}{R}$$

with the energy flux $\bar{I}$ interpreted as a power density and $|\mathbf{E}|^2$, proportional to the energy density as shown by Eq. 1.8, identified with the square of the voltage. It then becomes evident that the constant of proportionality can be regarded as $1/R$.

Problem 1.1 *Show that* $\sqrt{\frac{\mu_0}{\epsilon_0}}$ *has units of resistance and the numerical value is* 376.7 Ω.

1.2 Field Modes in a Cavity

We begin our discussion of light–matter interaction by establishing some basic ideas from the classical theory of radiation. What we seek to do is calculate the energy density inside a bounded conducting volume. We will then use this result to describe the interaction of the light with a collection of two-level atoms inside the cavity.

The basic physical idea is to consider that the electrons inside the conducting volume boundary oscillate as a result of thermal motion and, through dipole radiation, set up electromagnetic standing waves inside the cavity. Because the cavity walls are conducting, the electric field $\mathbf{E}$ must be zero there. Our task is twofold: first to count the number of standing waves that satisfy this boundary condition as a function of frequency; second, to assign an energy to each wave, and thereby determine the spectral distribution of energy density in the cavity.

The equations that describe the radiated energy in space are

$$\nabla^2\mathbf{E} = \frac{1}{c^2}\frac{\partial^2\mathbf{E}}{\partial t^2} \tag{1.13}$$

with

$$\nabla\cdot\mathbf{E} = 0 \tag{1.14}$$

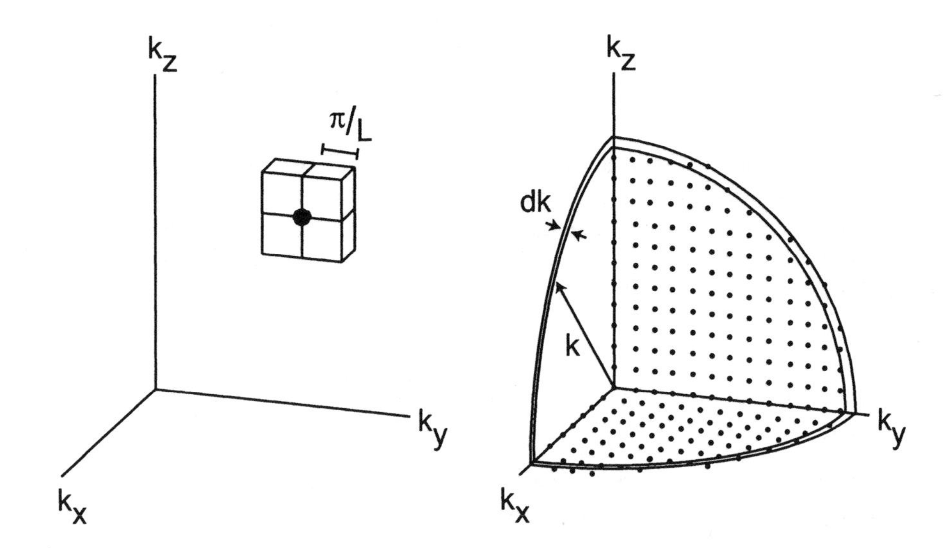

Figure 1.1: Mode points in **k** space. Left panel shows one-half the volume surrounding each point. Right panel shows one-eighth the volume of spherical shell in this **k** space.

Standing-wave solutions factor into oscillatory temporal and spatial terms. Now, respecting the boundary conditions for a three-dimensional box with sides of length L, we have for the components of $\mathbf{E}$

$$\begin{aligned} E_x(x,t) &= E_{0x}e^{-i\omega t}\cos(k_x x)\sin(k_y y)\sin(k_z z) \\ E_y(y,t) &= E_{0y}e^{-i\omega t}\sin(k_x x)\cos(k_y y)\sin(k_z z) \\ E_z(z,t) &= E_{0z}e^{-i\omega t}\sin(k_x x)\sin(k_y y)\cos(k_z z) \end{aligned} \tag{1.15}$$

where again $\mathbf{k}$ is the wave vector of the light, with amplitude

$$|\mathbf{k}| = \frac{2\pi}{\lambda} \tag{1.16}$$

and components

$$k_x = \frac{\pi n}{L} \quad n = 0, 1, 2, ... \tag{1.17}$$

and similarly for k_y, k_z. Notice that the cosine and sine factors for the E_x field component show that the transverse field amplitudes E_y, E_z have nodes at 0 and L as they should and similarly for E_y and E_z. In order to calculate the mode density, we begin by constructing a three-dimensional orthogonal lattice of points in $\mathbf{k}$ space as shown in Fig. 1.1. The separation between points along the k_x, k_y, k_z axes is $\frac{\pi}{L}$, and the volume associated with each point is therefore

$$V = \left(\frac{\pi}{L}\right)^3$$

Now the volume of a spherical shell of radius $|\mathbf{k}|$ and thickness dk in this space is $4\pi k^2 dk$. However, the periodic boundary conditions on $\mathbf{k}$ restrict k_x, k_y, k_z to positive values, so the effective shell volume lies only in the positive octant of the sphere. The number of points is therefore just this volume divided by the volume per point:

$$\text{Number of } k \text{ points in spherical shell} = \frac{\frac{1}{8}\left(4\pi k^2 dk\right)}{\left(\frac{\pi}{L}\right)^3} = \frac{1}{2}L^3\frac{k^2 dk}{\pi^2} \tag{1.18}$$

Remembering that there are two independent polarization directions per k point, we find that the number of radiation modes between k and dk is,

$$\text{Number of modes in spherical shell} = L^3\frac{k^2 dk}{\pi^2} \tag{1.19}$$

and the spatial density of modes in the spherical shell is

$$\frac{\text{Number of modes in shell}}{L^3} = d\rho(k) = \frac{k^2 dk}{\pi^2} \tag{1.20}$$

We can express the spectral mode density, mode density per unit k, as

$$\frac{d\rho(k)}{dk} = \rho_k = \frac{k^2}{\pi^2} \tag{1.21}$$

and therefore the mode number as

$$\rho_k dk = \frac{k^2}{\pi^2}dk \tag{1.22}$$

with ρ_k as the *mode density* in k-space. The expression for the mode density can be converted to frequency space, using the relations

$$k = \frac{2\pi}{\lambda} = \frac{2\pi\nu}{c} = \frac{\omega}{c} \tag{1.23}$$

and

$$\frac{d\nu}{dk} = \frac{c}{2\pi}$$

Clearly

$$\rho_\nu d\nu = \rho_k dk$$

and therefore

$$\rho_\nu d\nu = \frac{8\pi\nu^2 d\nu}{c^3} \tag{1.24}$$

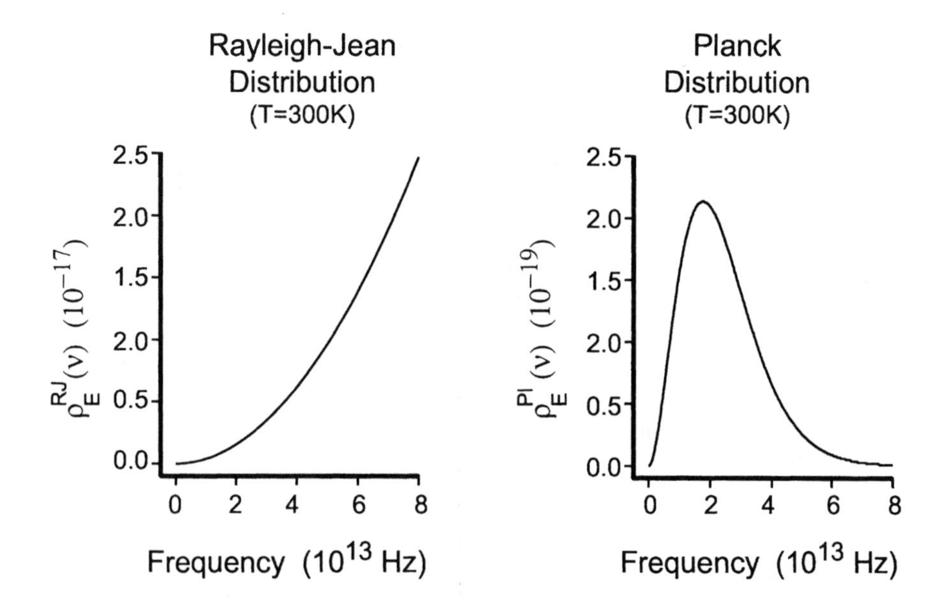

Figure 1.2: Left panel: Rayleigh–Jeans blackbody energy density distribution as a function of frequency, showing the rapid divergence as frequencies tend toward the ultraviolet (the ultraviolet catastrophe). Right panel: Planck blackbody energy density distribution showing correct high-frequency behavior.

The density of oscillator modes in the cavity increases as the square of the frequency. Now the average energy per mode of a collection of oscillators in thermal equilibrium, according to the principal of equipartition of energy, is equal to k_BT, where k_B is the Boltzmann constant. We conclude therefore that the *energy density* in the cavity is

$$\rho_E^{RJ}(\nu)d\nu = \frac{8\pi\nu^2 k_B T d\nu}{c^3} \tag{1.25}$$

which is known as the *Rayleigh–Jeans law of blackbody radiation*; and, as Fig. 1.2 shows, leads to the unphysical conclusion that energy storage in the cavity increases as the square of the frequency without limit. This result is sometimes called the "ultraviolet catastrophe" since the energy density increases without limit as oscillator frequency increases toward the ultraviolet region of the spectrum. We achieved this result by multiplying the number of modes in the cavity by the average energy per mode. Since there is nothing wrong with our mode counting, the problem must be in the use of the equipartition principle to assign energy to the oscillators.

1.2.1 Planck mode distribution

We can get around this problem by first considering the mode excitation probability distribution of a collection of oscillators in thermal equilibrium at temperature T. This probability distribution P_i comes from statistical mechanics and can be written in terms of the Boltzmann factor $e^{-\epsilon_i/k_BT}$ and the partition function $q = \sum_{i=0}^{\infty} e^{-\epsilon_i/k_BT}$:

$$P_i = \frac{e^{-\epsilon_i/k_BT}}{q}$$

Now, Planck suggested that instead of assigning the average energy k_BT to every oscillator, this energy could be assigned in discrete amounts, proportional to the frequency, such that

$$\epsilon_i = \left(n_i + \frac{1}{2}\right) h\nu$$

where $n_i = 0, 1, 2, 3\ldots$ and the constant of proportionality $h = 6.626 \times 10^{-34}$ J·s. We then have

$$P_i = \frac{e^{-h\nu/2k_BT} e^{-n_i h\nu/k_BT}}{e^{-h\nu/2k_BT} \sum_{n_i=0}^{\infty} e^{-n_i h\nu/k_BT}} = \frac{\left(e^{-h\nu/k_BT}\right)^{n_i}}{\sum_{n_i=0}^{\infty} \left(e^{-h\nu/k_BT}\right)^{n_i}} \tag{1.26}$$

$$= \left(e^{-h\nu/k_BT}\right)^{n_i} \left(1 - e^{-h\nu/k_BT}\right) \tag{1.27}$$

where we have recognized that $\sum_{n_i=0}^{\infty} \left(e^{-h\nu/k_BT}\right)^{n_i} = 1/\left(1 - e^{-h\nu/k_BT}\right)$. The average energy per mode then becomes

$$\bar{\varepsilon} = \sum_{i=0}^{\infty} P_i \epsilon_i =$$

$$\sum_{n_i=0}^{\infty} \left(e^{-h\nu/k_BT}\right)^{n_i} \left(1 - e^{-h\nu/k_BT}\right) (n_i) h\nu = \frac{h\nu}{e^{h\nu/k_BT} - 1} \tag{1.28}$$

and we obtain the Planck energy density in the cavity by substituting $\bar{\varepsilon}$ from Eq. 1.28 for k_BT in Eq. 1.25

$$\rho_E^{Pl}(\nu)d\nu = \frac{8\pi h}{c^3} \nu^3 \frac{1}{e^{h\nu/k_BT} - 1} d\nu \tag{1.29}$$

This result, plotted in Fig. 1.2, is much more satisfactory than the Rayleigh–Jeans result since the energy density has a bounded upper limit and the distribution agrees with experiment.

Problem 1.2 *Prove Eq. 1.28 using the closed form for the geometric series,* $\sum_{n_i=0}^{\infty}\left(e^{-h\nu/k_BT}\right)^{n_i}$ *and* $\frac{ds^{n_i}}{ds}=n_i s^{n_i-1}$, *where* $s=e^{-h\nu/k_BT}$.

Problem 1.3 *Show that Eq. 1.29 assumes the form of the Rayleigh-Jeans law (Eq. 1.25) in the low-frequency limit.*

1.3 The Einstein *A* and *B* coefficients

Let us consider a two-level atom or collection of atoms inside the conducting cavity. We have N_1 atoms in the lower level E_1 and N_2 atoms in the upper level E_2. Light interacts with these atoms through resonant stimulated absorption and emission, $E_2 - E_1 = \hbar\omega_0$, the rates of which, $B_{12}\rho_\omega$, $B_{21}\rho_\omega$, are proportional to the spectral energy density ρ_ω of the cavity modes. Atoms populated in the upper level can also emit light "spontaneously" at a rate A_{21} that depends only on the density of cavity modes (i.e., the volume of the cavity). This phenomenological description of light absorption and emission can be described by rate equations first written down by Einstein. These rate equations were meant to interpret measurements in which the spectral width of the radiation sources was broad compared to a typical atomic absorption line width and the source spectral flux $\bar{I}_\omega$ (W/m^2·Hz) was weak compared to the saturation intensity of a resonant atomic transition. Although modern laser sources are, according to these criteria, both narrow and intense, the spontaneous rate coefficient A_{21} and the stimulated absorption coefficient B_{12} are still often used in the spectroscopic literature to characterize light–matter interaction in atoms and molecules.

These Einstein rate equations describe the energy flow between the atoms in the cavity and the field modes of the cavity, assuming of course, that total energy is conserved:

$$\frac{dN_1}{dt} = -\frac{dN_2}{dt} = -N_1B_{12}\rho_\omega + N_2B_{21}\rho_\omega + N_2A_{21} \tag{1.30}$$

At thermal equilibrium we have a steady-state condition $\frac{dN_1}{dt} = -\frac{dN_2}{dt} = 0$ with $\rho_\omega = \rho_\omega^{\text{th}}$ so that

$$\rho_\omega^{\text{th}} = \frac{A_{21}}{\left(\frac{N_1}{N_2}\right)B_{12} - B_{21}}$$

and the Boltzmann distribution controlling the distribution of the number of atoms in the lower and upper levels,

$$\frac{N_1}{N_2} = \frac{g_1}{g_2}e^{-(E_1-E_2)/kT}$$

where g_1, g_2 are the degeneracies of the lower and upper states, respectively. So

$$\rho_\omega^{\text{th}} = \frac{A_{21}}{\left(\frac{g_1}{g_2}e^{\hbar\omega_0/kT}\right)B_{12} - B_{21}} = \frac{\frac{A_{21}}{B_{21}}}{\left(\frac{g_1}{g_2}e^{\hbar\omega_0/kT}\right)\frac{B_{12}}{B_{21}} - 1} \tag{1.31}$$

But this result has to be consistent with the Planck distribution, Eq. 1.29:

$$\rho_E^{Pl}(\nu)\,d\nu = \frac{8\pi h}{c^3}\nu_0^3\frac{1}{e^{h\nu_0/k_BT}-1}\,d\nu \tag{1.32}$$

$$\rho_E^{Pl}(\omega)\,d\omega = \frac{\hbar}{\pi^2c^3}\omega_0^3\frac{1}{e^{\hbar\omega_0/k_BT}-1}\,d\omega \tag{1.33}$$

Therefore, comparing these last two expressions with Eq. 1.31, we must have

$$\frac{g_1}{g_2}\frac{B_{12}}{B_{21}} = 1 \tag{1.34}$$

and

$$\frac{A_{21}}{B_{21}} = \frac{8\pi h}{c^3}\nu_0^3 \tag{1.35}$$

or

$$\frac{A_{21}}{B_{21}} = \frac{\hbar\omega_0^3}{\pi^2c^3} \tag{1.36}$$

These last two equations show that if we know one of the three rate coefficients, we can always determine the other two.

It is worthwhile to compare the spontaneous emission rate A_{21} to the stimulated emission rate B_{21}

$$\frac{A_{21}}{B_{21}\rho_\omega^{th}} = e^{\hbar\omega_0/kT} - 1$$

which shows that for $\hbar\omega_0$ much greater than kT (visible, UV, X-ray), the spontaneous emission rate dominates; but for regions of the spectrum much less than kT (far IR, microwaves, radio waves), the stimulated emission process is much more important. It is also worth mentioning that even when stimulated emission dominates, spontaneous emission is always present. We shall see in Appendix 7.B that in fact spontaneous emission "noise" is the ultimate factor limiting laser line narrowing.

1.4 Light Propagation in a Dielectric Medium

So far we have assumed that light propagates either through a vacuum or through a gas so dilute that we need consider only the isolated field–atom interaction. Now we consider the propagation of light through a continuous dielectric (nonconducting) medium. Interaction of light with such a medium permits us to introduce the important quantities of polarization, susceptibility, index of refraction, extinction coefficient, and absorption coefficient. We shall see later (Section 4.4.1 and Chapter 7) that the polarization can be usefully regarded as a density of transition dipoles induced in the dielectric by the oscillating light field, but here we begin by simply defining the polarization $\mathbf{P}$ with respect to an applied electric field $\mathbf{E}$ as

$$\mathbf{P} = \epsilon_0\chi\mathbf{E} \tag{1.37}$$

where χ is the linear electric *susceptibility*, an intrinsic property of the medium responding to the light field.

It is worthwhile to digress for a moment and recall the relation between the electric field $\mathbf{E}$, the polarization $\mathbf{P}$, and the displacement field $\mathbf{D}$ in a material medium. In the rationalized MKS system of units the relation is

$$\mathbf{D} = \epsilon_0 \mathbf{E} + \mathbf{P} \tag{1.38}$$

Furthermore, for isotropic materials, in all systems of units, the so-called constitutive relation between the displacement field $\mathbf{D}$ and the imposed electric field $\mathbf{E}$ is written

$$\mathbf{D} = \epsilon \mathbf{E}$$

with ϵ referred to as the *dielectric constant* of the material. Therefore

$$\mathbf{D} = \epsilon_0 (1 + \chi) \mathbf{E}$$

and

$$\epsilon = \epsilon_0 (1 + \chi)$$

The susceptibility χ is often a strong function of frequency ω around resonances and can be spatially anisotropic. It is a complex quantity having a real, dispersive component χ' and an imaginary absorptive component χ'':

$$\chi = \chi' + i\chi''$$

A number of familiar expressions in free space become modified in a dielectric medium. For example

$$\left(\frac{kc}{\omega}\right)^2 = 1 \qquad \text{; free space}$$

$$\left(\frac{kc}{\omega}\right)^2 = 1 + \chi \text{ ; dielectric}$$

In a dielectric medium $\frac{kc}{\omega}$ becomes a complex quantity that is conventionally expressed as

$$\frac{kc}{\omega} = \eta + i\kappa$$

where η is the *refractive index* and κ is the *extinction coefficient* of the dielectric medium. The relations between the refractive index, the extinction coefficient, and the two components of the susceptibility are

$$\eta^2 - \kappa^2 = 1 + \chi'$$
$$2\eta\kappa = \chi''$$

Note that in a transparent dielectric medium

$$\eta^2 = 1 + \chi' = \frac{\epsilon}{\epsilon_0} \tag{1.39}$$

In a dielectric medium the traveling wave solutions of Maxwell's equation become,

$$\mathbf{E} = \mathbf{E_0} e^{i(kz-\omega t)} \longrightarrow \mathbf{E_0} e^{\left[i\omega\left(\frac{\eta z}{c}-t\right)-\omega\frac{\kappa}{c}z\right]}$$

the relation between magnetic and electric field amplitudes is

$$B_0 = \sqrt{\epsilon_0\mu_0}\, E_0 \longrightarrow B_0 = \sqrt{\epsilon_0\mu_0}\,(\eta + i\kappa)\, E_0$$

and the period-averaged field energy density is

$$\bar{\rho}_\omega = \frac{1}{2}\epsilon_0 \left|\mathbf{E}\right|^2 \longrightarrow \bar{\rho}_\omega = \frac{1}{2}\epsilon_0\eta^2 \left|\mathbf{E}\right|^2 \tag{1.40}$$

Now the light-beam intensity in a dielectric medium is attenuated exponentially by absorption:

$$\bar{I}_\omega = \frac{1}{2}\epsilon_0 c \left|\mathbf{E}\right|^2 \longrightarrow \bar{I}_\omega = \frac{1}{2}\epsilon_0\eta^2 \left|\mathbf{E}\right|^2 \left(\frac{c}{\eta}\right) = \frac{1}{2}\epsilon_0 c\eta E_0^2 e^{-2\frac{\omega\kappa}{c}z} = \bar{I}_0 e^{-Kz} \tag{1.41}$$

where

$$\bar{I}_0 = \frac{1}{2}\epsilon_0 c\eta E_0^2 \tag{1.42}$$

is the intensity at the point where the light beam enters the medium, and

$$K = 2\frac{\omega\kappa}{c} = \frac{\omega}{\eta c}\chi'' \tag{1.43}$$

is termed the *absorption coefficient.* Note that the energy flux $\bar{I}_\omega$ in the dielectric medium is still the product of the energy density

$$\bar{\rho}_\omega = \frac{1}{2}\epsilon_0\eta^2 \left|\mathbf{E}\right|^2$$

and the speed of propagation c/η. Note also that, although light propagating through a dielectric maintains the same frequency as in vacuum, the wavelength contracts as

$$\lambda = \frac{c/\eta}{\nu}$$

1.5 Light Propagation in a Dilute Gas

We are often very interested in the attenuation of intensity as a light beam passes through a dilute gas of resonantly scattering atoms. Equation 1.41 describes this attenuation in terms of the material properties of a dielectric medium, but what we seek is an equivalent microscopic description in terms of the rate of atomic absorption and reemission of light. The Einstein rate equations tell us the time rate of absorption and emission, but what we would like to find is an expression that relates this *time* rate of change to a *spatial* rate of change along the light path. We consider a light beam propagating through a cell containing an absorbing gas and assume that, along the light-beam axis, absorption and reemission have reached steady state. We start with the expression for the Einstein rate equations, Eq. 1.30, and write

$$0 = -N_1 B_{12} \bar{\rho}_\omega + N_2 B_{21} \bar{\rho}_\omega + N_2 A_{21}$$

where $\bar{\rho}_\omega$ where the overbar on the energy density of the light beam indicates that it is averaged over a period of oscillation (see Eqs. 1.7, 1.8). We use the result from Eq. 1.34 to write

$$N_2 A_{21} = \bar{\rho}_\omega \left[N_1 B_{12} - N_2 B_{21}\right] = \bar{\rho}_\omega B_{12} \left[N_1 - N_2 \frac{g_1}{g_2}\right] \tag{1.44}$$

At steady state the number of excited atoms is

$$N_2 = \frac{\bar{\rho}_\omega B_{12} N_1}{A_{21} + \frac{g_1}{g_2} \bar{\rho}_\omega B_{12}} \tag{1.45}$$

Now, when considering propagation through a dilute gas, we have to be careful to take into account correctly the index refraction of the dielectric medium. The expression for the energy density ρ_ω in terms of the field energy and the cavity volume must be modified according to Eq. 1.40, so that

$$\bar{\rho}_\omega \text{ (vacuum)} \longrightarrow \bar{\rho}_\omega \eta^2 \text{ (dielectric)} \tag{1.46}$$

In order to use the Einstein rate coefficients, which assume propagation at the speed of light in vacuum, we have to "correct" the energy density ρ_ω in the dielectric medium before inserting it into Eq. 1.45. Therefore $\bar{\rho}_\omega$ in Eq. 1.45 must be replaced by $\bar{\rho}_\omega / \eta^2$:

$$N_2 A_{21} = \frac{\bar{\rho}_\omega}{\eta^2} B_{12} \left[N_1 - N_2 \frac{g_1}{g_2}\right] \tag{1.47}$$

If we multiply both sides of Eq. 1.47 by $\hbar\omega_0$, the left-hand side describes the rate of energy scattered out of light beam in spontaneous emission

$$N_2 A_{21} \hbar\omega_0 \tag{1.48}$$

and the right side describes the net energy loss from the beam, that is, the difference between the energy removed by stimulated absorption and the energy returned to the beam by stimulated emission:

$$\frac{d\bar{U}_\omega}{dt} = -\frac{\bar{\rho}_\omega}{\eta^2} B_{12} \left[N_1 - N_2 \frac{g_1}{g_2}\right] \hbar\omega_0 \tag{1.49}$$

1.5.1 Spectral line shapes

Light sources always have an associated spectral width. Conventional light sources such as incandescent lamps or plasmas are broadband relative to atomic or molecular absorbers, at least in the dilute gas phase. Even if we use a very pure spectral source, like a laser tuned to the peak of an atomic resonance at ω_0, atomic transition lines always exhibit an intrinsic spectral width associated with an interruption of the phase evolution in the excited state. Phase interruptions such as spontaneous emission, stimulated emission, and collisions are common examples of such line broadening phenomena. The emission or absorption of radiation actually occurs over a distribution of frequencies centered on ω_0, and we have to take into account this spectral distribution in our energy balance. Rather than using Eq. 1.48, we more realistically express the rate of energy loss by spontaneous emission as

$$N_2 A_{21} \hbar \int_{\omega_0 - \frac{\Delta\omega}{2}}^{\omega_0 + \frac{\Delta\omega}{2}} \omega L\left(\omega - \omega_0\right) d\omega$$

where $L\left(\omega - \omega_0\right)$ is the atomic absorption line-shape function, usually normalized such that $\int L\left(\omega - \omega_0\right) d\omega = 1$. A common line-shape function in atomic spectroscopy is the Lorentzian,

$$L(\omega - \omega_0) d\omega = \frac{A_{21}}{2\pi} \frac{d\omega}{\left(\omega - \omega_0\right)^2 + \left(\frac{A_{21}}{2}\right)^2}$$

with spectral width equal to A_{21}. The differential $L(\omega - \omega_0)d\omega$ can be regarded as the probability of finding light emitted in the frequency interval between ω and $\omega + d\omega$. In fact, we shall encounter the Lorentzian line shape again when we consider other contributions to the spectral width such as strong-field excitation (power broadening) or collision broadening. More generally, therefore, we can write the normalized Lorentzian line-shape function as

$$L(\omega - \omega_0) d\omega = \frac{\gamma'}{2\pi} \frac{d\omega}{\left(\omega - \omega_0\right)^2 + \left(\frac{\gamma'}{2}\right)^2} \tag{1.50}$$

where γ' may be a composite of several physical sources for the spectral line width. Sometimes we are more interested in the *spectral density distribution function*, which is simply the line-shape function without normalization. For an atomic line broadened to a width γ', we obtain

$$F(\omega - \omega_0) d\omega = \frac{d\omega}{\left(\omega - \omega_0\right)^2 + \left(\frac{\gamma'}{2}\right)^2}$$

Note that $L(\omega - \omega_0)d\omega$ is unitless while $F(\omega - \omega_0)d\omega$ has units of the reciprocal of angular frequency or (2π) Hz^{-1}. Now, with $\bar{\rho}_\omega = d\bar{\rho}(\omega)/d\omega$, and generalizing Eq. 1.49, the corresponding net energy loss from the light beam is:

$$\frac{d\bar{U}}{dt} = -\int \frac{d\bar{\rho}(\omega')}{d\omega'} \frac{B_{12}}{\eta^2} \left[N_1 - N_2 \frac{g_1}{g_2}\right] \hbar\omega L\left(\omega - \omega_0\right) d\omega' d\omega$$

where the integral over ω' takes into account the spectral width of the light source and the *energy density* attenuation is

$$\frac{d\bar{\rho}}{dt} = -\int \frac{d\bar{\rho}(\omega')}{d\omega'} \frac{B_{12}}{\eta^2 V} \left[N_1 - N_2 \frac{g_1}{g_2}\right] \hbar\omega L(\omega - \omega_0)\, d\omega' d\omega \tag{1.51}$$

where V is the cavity volume. If we assume that the light beam propagates along z and convert the time dependence to a space dependence

$$\frac{d\rho}{dt} = \frac{d\rho}{dz} \cdot c = \frac{dI}{dz} \tag{1.52}$$

then, from inspection of Eqs. 1.40 and 1.41, we see that

$$I = \rho \frac{c}{\eta} \tag{1.53}$$

and substituting Eqs. 1.53 and 1.52 into 1.51, we finally obtain

$$\frac{d\bar{I}}{dz} = -\int_{\omega_0 - \frac{\Delta\omega}{2}}^{\omega_0 + \frac{\Delta\omega}{2}} \frac{dI(\omega')}{d\omega'} \frac{B_{12}}{c\eta V} \left[N_1 - N_2 \frac{g_1}{g_2}\right] \hbar\omega F(\omega - \omega_0)\, d\omega' d\omega \tag{1.54}$$

Now if the light only weakly excites the gas so that $N_2 << N_1$, we have

$$\frac{d\bar{I}}{dz} = -\int_{\omega_0 - \frac{\Delta\omega}{2}}^{\omega_0 + \frac{\Delta\omega}{2}} \frac{dI(\omega')}{d\omega'} \frac{B_{12}}{c\eta} n\hbar\omega F(\omega - \omega_0)\, d\omega' d\omega \tag{1.55}$$

where $n = N/V \simeq N_1/V$ is the gas density. In a weak light field and a dilute gas, we can obtain a simple expression for the intensity behavior by approximating the spectral distribution of the absorption with a Lorentzian spectral distribution function peaked at ω_0 with a width A_{21}:

$$\int_{\omega_0 - \frac{\Delta\omega}{2}}^{\omega_0 + \frac{\Delta\omega}{2}} \hbar\omega F(\omega - \omega_0)\, d\omega \simeq \hbar\omega_0 \int_{-\infty}^{+\infty} \frac{1}{(\omega - \omega_0)^2 + \left(\frac{A_{21}}{2}\right)^2} d\omega = \hbar\omega_0 \frac{2\pi}{A_{21}}$$

Then Eq. 1.55 becomes, with some rearrangement

$$\frac{d\bar{I}}{\bar{I}} = -\left[\frac{B_{12}}{A_{21}} \frac{2\pi\hbar\omega_0}{c\eta}\right] n dz \tag{1.56}$$

The term on the right in brackets has units of area and can be regarded as an expression for the cross section of absorption of resonant light. Using Eq. 1.36, we can express this cross section as

$$\sigma_0 = \frac{g_2}{g_1} \frac{\pi\lambda_0^2}{2\eta} \tag{1.57}$$

so that Eq. 1.56 can be written as

$$\frac{d\bar{I}}{\bar{I}} = -\sigma_0 n dz$$

and

$$\frac{\bar{I}}{\bar{I}_0} = e^{-\sigma_0 n z_0} \tag{1.58}$$

where z_0 is the total distance over which the absorption takes place. Equation 1.58 is the familiar integral form of the Lambert–Beer law for light absorption. It is quite useful for measuring atom densities in gas cells or beams. Comparing Eqs. 1.43 and 1.58, we see that the absorption coefficient K can be written as the product of the absorption cross section and the gas density:

$$K = \frac{2\omega\kappa}{c} = \frac{\omega}{\eta c}\chi'' = \sigma_0 n \tag{1.59}$$

Problem 1.4 *Suppose that a light beam enters a gas-filled cell with intensity I_0 at position z_0. Show that at high power such that $B_{21}\bar{\rho}_\omega \gg A_{21}$, the intensity in the beam decreases linearly with distance such that*

$$I - I_0 = -\frac{g_1}{g_2} n A_{21} \hbar \left[\int_{\omega_0 - \frac{\Delta\omega}{2}}^{\omega_0 + \frac{\Delta\omega}{2}} \omega L(\omega) d\omega \right] (z - z_0)$$

1.6 Further Reading

A thorough discussion of the various systems of units in electricity and magnetism can be found in

- J. D. Jackson *Classical Electrodynamics*, Wiley, New York, 1962. This book is currently in its third edition.

For the approach to mode counting in a conducting cavity and to light propagation in a dilute dielectric medium, we have followed

- R. Louden, *The Quantum Theory of Light,* 2nd edition, chapter 1, Clarendon Press, Oxford, 1983. This book is currently in its third edition.

Early history of the quantum theory and the problem of blackbody radiation may be found in

- J. C. Slater, *Quantum Theory of Atomic Structure* vol.I, chapter 1, McGraw-Hill, New York, 1960

Chapter 2

Semiclassical Treatment of Absorption and Emission

2.1 Introduction

In the previous chapter we introduced the Einstein A and B coefficients and associated them with the Planck spectral distribution of blackbody radiation. This procedure allowed us to relate the spontaneous and stimulated rate coefficients, but it did not provide any means to calculate them from intrinsic atomic properties. The goal of the present chapter is to find expressions for the rate of atomic absorption and emission of radiation from quantum mechanics and to relate these expression to the Einstein coefficients. As for all physical observables, we will find that these rates must be expressed in terms of probabilities of absorption and emission. Various disciplines such as spectrometry, spectroscopy, and astrophysics have developed their own terminologies to express these absorption and emission properties of matter, and we shall point out how many commonly encountered parameters are related to the fundamental transition probabilities and to each other. We restrict the discussion to the simplest of all structures: the two-level, nondegenerate, spinless atom.

2.2 Coupled Equations of the Two-Level System

We start with the time-dependent Schrödinger equation

$$\hat{H}\Psi\left(\mathbf{r},t\right)=i\hbar\frac{d\Psi}{dt} \tag{2.1}$$

and write the stationary-state solution of level n as

$$\Psi_n\left(\mathbf{r},t\right)=e^{-iE_nt/\hbar}\psi_n\left(\mathbf{r}\right)=e^{-i\omega_nt}\psi_n\left(\mathbf{r}\right)$$

The time-independent Schrödinger equation then becomes

$$\hat{H}_A\psi_n\left(\mathbf{r}\right)=E_n\psi_n\left(\mathbf{r}\right)$$

where the subscript A indicates "atom". Then for the two-level system we have

$$\hat{H}_A \psi_1 = E_1 \psi_1 = \hbar\omega_1 \psi_1$$
$$\hat{H}_A \psi_2 = E_2 \psi_2 = \hbar\omega_2 \psi_2$$

and write

$$\hbar\omega_0 = \hbar\left(\omega_2 - \omega_1\right) = E_2 - E_1$$

Now we add a time-dependent term to the Hamiltonian that will turn out to be proportional to the oscillating classical field with frequency not far from ω_0:

$$\hat{H} = \hat{H}_A + \hat{V}(t) \tag{2.2}$$

With the field turned on, the state of the system becomes a time-dependent linear combination of the two stationary states

$$\Psi\left(\mathbf{r}, t\right) = C_1\left(t\right)\psi_1 e^{-i\omega_1 t} + C_2\left(t\right)\psi_2 e^{-i\omega_2 t} \tag{2.3}$$

which we require to be normalized:

$$\int \left|\Psi\left(\mathbf{r}, t\right)\right|^2 d\tau = \left|C_1(t)\right|^2 + \left|C_2(t)\right|^2 = 1$$

Now, if we substitute the time-dependent wavefunction (Eq. 2.3) back into the time-dependent Schrödinger equation (Eq. 2.1), multiply on the left with $\psi_1^* e^{i\omega_1 t}$, and integrate over all space, we get

$$C_1 \int \psi_1^* \hat{V} \psi_1 d\mathbf{r} + C_2 e^{-i\omega_0 t} \int \psi_1^* \hat{V} \psi_2 d\mathbf{r} = i\hbar \frac{dC_1}{dt}$$

From now on we will denote the matrix elements $\int \psi_1^* \hat{V} \psi_1 d\mathbf{r}$ and $\int \psi_1^* \hat{V} \psi_2 d\mathbf{r}$ as V_{11} and V_{12}, so we have

$$C_1 V_{11} + C_2 e^{-i\omega_0 t} V_{12} = i\hbar \frac{dC_1}{dt} \tag{2.4}$$

and similarly for C_2, we obtain

$$C_1 e^{i\omega_0 t} V_{21} + C_2 V_{22} = i\hbar \frac{dC_2}{dt} \tag{2.5}$$

These two coupled equations define the quantum-mechanical problem and their solutions, C_1 and C_2, define the time evolution of the state wavefunction, Eq. 2.3. Of course, any measurable quantity is related to $\left|\Psi\left(\mathbf{r}, t\right)\right|^2$; consequently we are really more interested in $\left|C_1\right|^2$ and $\left|C_2\right|^2$ then the coefficients themselves.

2.2.1 Field coupling operator

A single-mode radiation source, such as a laser, aligned along the z-axis, will produce an electromagnetic wave with amplitude E_0, polarization $\hat{\mathbf{e}}$, and frequency ω

$$\mathbf{E} = \hat{\mathbf{e}} E_0 \cos\left(\omega t - kz\right)$$

with the magnitude of the wave vector, as expressed in Eq. 1.23:

$$k = \frac{2\pi}{\lambda} \qquad \text{and} \qquad \omega = 2\pi\nu = 2\pi\frac{c}{\lambda} = kc$$

Now if we take a typical optical wavelength in the visible region of the spectrum, say, $\lambda = 600$ nm $\simeq 11{,}000\ a_0$, it is clear that these wavelengths are much longer than the characteristic length of an atom ($\simeq a_0$). Therefore over the spatial extent of the interaction between the atom and the field the kz term ($\simeq ka_0$) in the cosine argument will be negligible, and we can consider the field to be constant in amplitude over the scale length of the atom. We can make the *dipole approximation* in which the leading interaction term between the atom and the optical field is the scalar product of the instantaneous atom dipole $\mathbf{d}$, defined as

$$\mathbf{d} = -e\mathbf{r} = -e\sum_j \mathbf{r_j} \tag{2.6}$$

(where the $\mathbf{r_j}$ are the radii of the various electrons in the atom), and the electric field $\mathbf{E}$ in Eq. 2.6 defines a classical dipole. The corresponding quantum-mechanical operator is

$$\hat{\mathbf{d}} = -e\hat{\mathbf{r}} = -e\sum_j \hat{\mathbf{r_j}}$$

and

$$\hat{V} = -\hat{\mathbf{d}} \cdot \mathbf{E}$$

Note that the operator $\hat{V}$ has odd parity with respect to the electron coordinate $\mathbf{r}$ so that matrix elements V_{11} and V_{22} must necessarily vanish, and only atomic states of opposite parity can be coupled by the dipole interaction. The explicit expression for V_{12} is

$$V_{12} = eE_0 r_{12} \cos\omega t$$

with

$$r_{12} = \int \psi_1^* \left(\sum_j \hat{\mathbf{r_j}} \cdot \hat{\mathbf{e}} \right) \psi_2 \, d\tau \tag{2.7}$$

The transition dipole moment matrix element is defined as

$$\mu_{12} \equiv er_{12} \tag{2.8}$$

Equation 2.7 describes the resultant electronic coordinate vector summed over all electrons and projected onto the electric field direction of the optical wave. It is convenient to collect all these scalar quantities into one term:

$$\Omega_0 = \frac{\mu_{12}E_0}{\hbar} = \frac{eE_0 r_{12}}{\hbar} \tag{2.9}$$

So finally we have

$$V_{12} = \hbar\Omega_0 \cos\omega t$$

2.2.2 Calculation of the Einstein B_{12} coefficient

Now we can go back to our coupled equations, Eqs. 2.4 and 2.5, and write them as

$$\Omega_0 \cos \omega t\, e^{-i\omega_0 t} C_2 = i\frac{dC_1}{dt} \tag{2.10}$$

and

$$\Omega_0^* \cos \omega t\, e^{i\omega_0 t} C_1 = i\frac{dC_2}{dt} \tag{2.11}$$

We take the initial conditions to be $C_1(t = 0) = 1$ and $C_2(t = 0) = 0$ and remember that $|C_2(t)|^2$ expresses the probability of finding the population in the excited state at time t. Now the time rate of increase for the probability of finding the atom in its excited state is given by

$$\frac{|C_2(t)|^2}{t}$$

but the excitation rate described by the phenomenological Einstein expression (Eq. 1.30) is given just by

$$B_{12}\rho_\omega\, d\omega$$

To find the link between the Einstein B coefficient and V_{12}, we equate the two quantities

$$B_{12}\rho(\omega)\, d\omega = \frac{|C_2(t)|^2}{t} \tag{2.12}$$

and seek the solution $C_2(t)$ from Eq. 2.11 and the initial conditions. In the weak-field regime where only terms linear in Ω_0 are important, we have

$$C_2(t) = \frac{\Omega_0^*}{2}\left[\frac{1 - e^{i(\omega_0+\omega)t}}{\omega_0 + \omega} + \frac{1 - e^{i(\omega_0-\omega)t}}{\omega_0 - \omega}\right] \tag{2.13}$$

If the frequency of the driving wave ω approaches the transition resonant frequency ω_0, the exponential in the first term in brackets will oscillate at about twice the atomic resonant frequency ω_0 ($\sim 10^{15}\ \mathrm{s}^{-1}$), very fast compared to the characteristic rate of weak-field optical coupling ($\sim 10^{8}\ \mathrm{s}^{-1}$). Therefore over the time of the transition, the first term in Eq. 2.13 will be negligible compared to the second. To a quite good approximation we can write,

$$C_2(t) \simeq \frac{\Omega_0^*}{2}\left[\frac{1 - e^{i(\omega_0-\omega)t}}{\omega_0 - \omega}\right] \tag{2.14}$$

The expression for $C_2(t)$ in Eq. 2.14 is called the *rotating wave approximation* (RWA). We now have

$$|C_2(t)|^2 = |\Omega_0|^2 \frac{\sin^2\left[(\omega_0 - \omega)\frac{t}{2}\right]}{(\omega_0 - \omega)^2} \tag{2.15}$$

and when $\omega \rightarrow \omega_0$, application of L'Hôpital's rule yields

$$|C_2(t)|^2 = \frac{1}{4} |\Omega_0|^2 t^2$$

Once again, in order to arrive at a practical expression relating $|C_2(t)|^2$ to the Einstein B coefficients, we have to take into account the fact that there is always a finite width in the spectral distribution of the excitation source. The source might be, for example, a "broadband" arc lamp or the output from a monochromator coupled to a synchrotron, or a "narrowband" monomode laser whose spectral width would probably be narrower than the natural width of the atomic transition. So if we write the field energy as an integral over the spectral energy density of the excitation source in the neighborhood of the transition frequency

$$\frac{1}{2}\epsilon_0 E_0^2 = \int_{\omega_0 - \frac{1}{2}\Delta\omega}^{\omega_0 + \frac{1}{2}\Delta\omega} \rho_\omega d\omega \tag{2.16}$$

where the limits of integration, $\omega_0 \pm \frac{1}{2}\Delta\omega$, refer to the spectral width of the excitation source, and recognize from Eq. 2.15 that

$$|C_2(t)|^2 = \left(\frac{eE_0 r_{12}}{\hbar}\right)^2 \frac{\sin^2\left[(\omega_0 - \omega)\frac{t}{2}\right]}{(\omega_0 - \omega)^2} \tag{2.17}$$

we can then substitute Eq. 2.16 into Eq. 2.17 to find

$$|C_2(t)|^2 = \frac{e^2 2 r_{12}^2}{\epsilon_0 \hbar^2} \int_{\omega_0 - \frac{1}{2}\Delta\omega}^{\omega_0 + \frac{1}{2}\Delta\omega} \rho_\omega \frac{\sin^2\left[(\omega_0 - \omega)\frac{t}{2}\right]}{(\omega_0 - \omega)^2} d\omega$$

For conventional "broadband" excitation sources we can safely assume that the spectral density is constant over the line width of the atomic transition and take $\rho_\omega d\omega$ outside the integral operation and set it equal to $\rho(\omega_0)$. Note that this approximation is *not* valid for narrowband monomode lasers. Let us assume a fairly broadband continuous excitation so that $t(\omega_0 - \omega) >> 1$. In this case

$$\int_{\omega_0 - \frac{1}{2}\Delta\omega}^{\omega_0 + \frac{1}{2}\Delta\omega} \frac{\sin^2\left[(\omega_0 - \omega)\frac{t}{2}\right]}{(\omega_0 - \omega)^2} d\omega = \frac{\pi t}{2}$$

and the expression for the probability of finding the atom in the excited state becomes

$$|C_2(t)|^2 = \frac{e^2 \pi r_{12}^2}{\epsilon_0 \hbar^2} \rho(\omega_0) t \tag{2.18}$$

Remembering that Eq. 2.12 provides the bridge between the quantum-mechanical and classical expressions for the rate of excitation, we now can write the Einstein

B coefficient in terms of the quantum-mechanical transition moment

$$B_{12}\rho(\omega_0) = \frac{|C_2(t)|^2}{t} = \frac{e^2 \pi r_{12}^2}{\epsilon_0 \hbar^2}\,\rho(\omega_0) \tag{2.19}$$

or

$$B_{12} = \frac{e^2 \pi r_{12}^2}{\epsilon_0 \hbar^2}$$

Now only two details remain to obtain the final result: first, assuming that the atoms move randomly within a confined space, we have to average the orientation of the dipole moment over all spatial directions with respect to the light-field polarization. Equation 2.7 defined r_{12} to be the projection of the transition moment in the same direction as the electric field polarization. Second, in real atoms ground and excited levels often have several degenerate states associated with them, so we have to take into account the degeneracies g_1 and g_2 of the lower and upper levels, respectively. The value of r_{12}^2 averaged over all angles of orientation is simply

$$\left\langle |r_{12}|^2 \right\rangle = r_{12}^2 \left\langle \cos^2\theta \right\rangle = \frac{1}{3} r_{12}^2$$

so we have finally

$$B_{12} = \frac{e^2 \pi r_{12}^2}{3\epsilon_0 \hbar^2} \tag{2.20}$$

or in terms of the matrix element of the transition moment, from Eq. 2.8:

$$B_{12} = \frac{\pi \mu_{12}^2}{3\epsilon_0 \hbar^2} \tag{2.21}$$

Furthermore we know that the Einstein B coefficient for stimulated emission is related to the coefficient for absorption by

$$g_1 B_{12} = g_2 B_{21}$$

so that

$$B_{21} = \frac{g_1}{g_2} B_{12} = \frac{g_1}{g_2} \frac{\pi \mu_{12}^2}{3\epsilon_0 \hbar^2}$$

and we also have the important expression from Eq. 1.36:

$$A_{21} = \frac{g_1}{g_2} \frac{\omega_0^3 \mu_{12}^2}{3\pi \epsilon_0 \hbar c^3} \tag{2.22}$$

Thus the expressions for the rates of absorption and stimulated and spontaneous emission are all simply related in terms of universal physical constants, the transition frequency ω_0, and μ_{12}.

2.2.3 Relations between transition moments, line strength, oscillator strength, and cross section

In addition to the Einstein coefficients A_{21}, B_{21}, B_{12}, the transition dipole moment amplitude μ_{12}, and the absorption cross section $\sigma_{0a}(\omega)$, three other quantities, the oscillator strength f, the line strength S, and the spectral absorption cross section σ_ω are sometimes used to characterize atomic transitions.

2.2.4 Line strength

The line strength S is defined as the square of the transition dipole moment summed over all degeneracies in the lower and upper levels:

$$S_{12} = S_{21} = \sum_{m_1, m_2} \left| \langle \psi_{1,m_1} | \mu | \psi_{2,m_2} \rangle \right|^2 \tag{2.23}$$

The line strength becomes meaningful when we have to deal with real atoms degenerate in the upper and lower levels. In such cases we have to extend our idea of μ_{12} to consider the individual transition dipole matrix elements between each degenerate sublevel of the upper and lower levels. For a nondegenerate two-level atom, the μ_{12} and A_{21} are simply related:

$$A_{21} = \frac{\omega_0^3}{3\pi\epsilon_0 \hbar c^3} \mu_{12}^2 \tag{2.24}$$

If the lower level were degenerate, calculation of the rate coefficient for spontaneous emission would include the summation over all possible downward radiative transitions. In this case μ_{12}^2 is defined as the sum of the coupling matrix elements between the upper state and all allowed lower states:

$$\mu_{12}^2 = \sum_{m_1} \left| \langle \psi_{1,m_1} | \mu | \psi_2 \rangle \right|^2 \tag{2.25}$$

Now it can be shown that the rate of spontaneous emission from any sublevel of a degenerate excited level to a lower level (i.e. the sum over all the lower sublevels), is the same for all the excited sublevels.[1] This statement reflects the intuitively plausible idea that spontaneous emission should be spatially isotropic and unpolarized if excited-state sublevels are uniformly populated. Therefore insertion of μ_{12}^2 from Eq. 2.25 in Eq. 2.24 would produce the correct result even if the upper level were degenerate. However, it would be tidier and notationally more symmetric to define a μ_{12}^2 summed over both upper and lower degeneracies:

$$\bar{\mu}_{12}^2 = \sum_{m_1 m_2} \left| \langle \psi_{1,m_1} | \mu | \psi_{2,m_2} \rangle \right|^2 \tag{2.26}$$

[1] The rate of spontaneous emission from multilevel atoms is properly outside the scope of a discussion of the two-level atom. The properties of spontaneous and stimulated emission are usually developed by expanding the transition moment in terms of spherical tensors and the atom wavefunctions in a basis of angular momentum states. One can then make use of angular momentum algebra, such as the $3j$ symbols, to prove that spontaneous emission is spatially isotropic and unpolarized.

Insertion of $\bar{\mu}_{12}^2$ from Eq. 2.26 in Eq. 2.24 must therefore be accompanied by a factor $\frac{1}{g_2}$ to correct for the fact that all excited sublevels radiate at the same rate. So, with $\bar{\mu}_{12}^2$ defined as in Eq. 2.26, the correct expression relating the transition dipole between degenerate levels to spontaneous emission rate becomes

$$A_{21} = \frac{1}{g_2} \frac{\omega_0^3}{3\epsilon_0 \hbar c^3} \bar{\mu}_{12}^2$$

The line strength defined in Eq. 2.23 is therefore related to A_{21} by

$$S_{12} = S_{21} = g_2 \frac{3\epsilon_0 \hbar c^3}{\omega_0^3} A_{21}$$

2.2.5 Oscillator strength

For an atom with two levels separated in energy by $\hbar\omega_0$, the *emission oscillator strength* is defined as a measure of the rate of radiative decay A_{21} compared to the radiative decay rate γ_e of a classical electron oscillator at ω_0 :

$$f_{21} = -\frac{1}{3} \frac{A_{21}}{\gamma_e}$$

In the case of degeneracies, the *absorption oscillator strength* is then defined as

$$f_{12} = -\frac{g_2}{g_1} f_{21} = \frac{g_2}{3g_1} \frac{A_{21}}{\gamma_e}$$

In real atoms, $S \longleftrightarrow P$ transitions behave approximately as classical oscillators; and the factor of $\frac{1}{3}$ in the definition compensates for the threefold degeneracy of P levels. Thus an $S \longleftrightarrow P$ transition, behaving exactly as a classical oscillator, would be characterized by an emission oscillator strength of $f_{21} = -\frac{1}{3}$ and an absorption oscillator strength $f_{12} = 1$. The classical expression for γ_e is

$$\gamma_e = \frac{e^2 \omega_0^2}{6\pi\epsilon_0 m_e c^3}$$

so in terms of the A_{21} coefficient and fundamental constants, the absorption oscillator strength is given by

$$f_{12} = A_{21} \frac{2\pi\epsilon_0 m_e c^3}{e^2 \omega_0^2}$$

Oscillator strengths obey certain sum rules that are useful in analyzing the relative intensities of atomic spectral lines. For example, one-electron atoms obey the following sum rule

$$\sum_k f_{ik} = 1 \tag{2.27}$$

where the summation is over all excited states, starting from the ground state. Alkali atoms are approximately one-electron systems, and the oscillator strength of the first $S \rightarrow P$ transition is typically on the order of 0.7–0.95. The sum rule tells us that most of the total transition probability for excitation of the valence electron is concentrated in the first $S \rightarrow P$ transition and that transitions to higher levels will be comparatively much weaker. Another sum rule exists for excitation and spontaneous emission from intermediate excited states j:

$$\sum_{i<j} f_{ji} + \sum_{k>j} f_{jk} = 1 \tag{2.28}$$

If the atomic spectrum can be ascribed to the motion of z electrons, then Eq. 2.28 can be generalized to

$$\sum_{i<j} f_{ji} + \sum_{k>j} f_{jk} = Z \tag{2.29}$$

which is called the *Thomas-Reiche-Kuhn sum rule*. In the multielectron form (Eq. 2.29) this sum rule is most useful when Z is the number of *equivalent* electrons, that is, electrons with the same n, l quantum numbers. Note also that the f_{ji} terms are intrinsically negative. Oscillator strengths are often used in astrophysics and plasma spectroscopy. They are sometimes tabulated as $\log gf$, where

$$g_1 f_{12} = -g_2 f_{21} \equiv gf$$

2.2.6 Cross section

The spectral absorption cross section σ_ω is associated with a beam of light propagating through a medium that absorbs and scatters the light by spontaneous emission. It is simply the ratio of absorbed power to propagating flux in the frequency interval between ω and $\omega + d\omega$:

$$\sigma_\omega = \frac{P(\omega)}{I(\omega)} \tag{2.30}$$

From Eqs. 1.34 and 1.36 we write

$$\begin{aligned} P(\omega) &= \hbar\omega B_{12}\rho_\omega \\ &= \hbar\omega \frac{g_2}{g_1} \frac{\pi^2 c^3}{\hbar\omega^3} A_{21}\rho_\omega \end{aligned} \tag{2.31}$$

and from Eqs. 1.10 and 1.11

$$I(\omega) = c\rho_\omega$$

so the "spectral" cross section, which has units of the product of area and frequency (e.g., $m^2 \cdot s^{-1}$) is

$$\sigma_\omega = \frac{g_2}{g_1} \frac{\pi^2 c^2}{\omega^2} A_{21} \tag{2.32}$$

from which we recover the "real" absorption cross section $\sigma_{0a}(\omega)$ (with units of area) by multiplying σ_ω by a line shape function $L(\omega - \omega_0)$. The subscript

a denotes absorption. Assuming a normalized Lorentzian lineshape function with width A_{21}

$$L(\omega - \omega_0) = \frac{A_{21}}{2\pi} \frac{1}{(\omega - \omega_0)^2 + \left(\frac{A_{21}}{2}\right)^2} \tag{2.33}$$

and replacing ω by ω_0 in Eq. 2.32, we obtain

$$\sigma_{0a}(\omega) = \frac{g_2}{g_1} \frac{c^2}{\omega_0^2} \frac{A_{21}^2}{2} \frac{1}{(\omega - \omega_0)^2 + \left(\frac{A_{21}}{2}\right)^2} \tag{2.34}$$

The substitution of ω_0 for ω is justified because the spectral cross section is sharply peaked around ω_0. The total absorption cross section, appropriate to broadband excitation covering the entire line profile, is obtained from multiplying σ_ω in Eq. 2.32 by the spectral distribution function $F(\omega - \omega_0)$ and integrating over the spectral width:

$$\sigma_{0a} = \int_{-\infty}^{+\infty} \frac{g_2}{g_1} \frac{\pi^2 c^2}{\omega_0^2} A_{21} F(\omega - \omega_0)\, d\omega =$$

$$\frac{g_2}{g_1} \frac{\pi^2 c^2}{\omega_0^2} A_{21} \int_{-\infty}^{+\infty} \frac{d\omega}{(\omega - \omega_0)^2 + \left(\frac{A_{21}}{2}\right)^2} \tag{2.35}$$

The result is

$$\sigma_{0a} = \frac{g_2}{g_1} \frac{2\pi^3 c^2}{\omega_0^2} = \frac{g_2}{g_1} \frac{\pi \lambda_0^2}{2} \tag{2.36}$$

consistent with Eq. 1.57. One obtains the *emission* cross section by substituting $B_{21} = \frac{g_1}{g_2} B_{12}$ for B_{12} in Eq. 2.31:

$$\sigma_{0e} = \frac{\pi^2 c^2}{\omega_0^2} = \frac{g_1}{g_2} \sigma_{0a}$$

Table 2.1 summarizes the various relations among these quantities used to characterized the absorption and emission of radiation. The quantity in the leftmost column is equal to the entry multiplied by the quantity in the topmost column.

Table 2.1: Conversion factors between Einstein A, B coefficients, transition dipole moment, oscillator strength, line strength, and cross section. The quantity in the leftmost column is equal to the entry multiplied by the quantity in the topmost column. Note that quantities refer to nondegenerate two-state transitions. Degeneracies of upper and lower levels are indicated by g_1 and g_2, respectively. Note also that σ_{ω_0} is the "spectral" cross section

	A_{21}	B_{12}	B_{21}	μ_{12}^2	f_{12}	f_{21}	S_{12}	σ_{ω_0}
A_{21}	1							
B_{12}	$\frac{g_2}{g_1}\frac{\pi^2 c^3}{\hbar\omega_0^3}$	1						
B_{21}	$\frac{\pi c^3}{\hbar\omega_0^3}$	$\frac{g_1}{g_2}$	1					
μ_{12}^2	$\frac{g_2}{g_1}\frac{3\pi\epsilon_0\hbar c^3}{\omega_0^3}$	$\frac{g_1}{g_2}\frac{3\epsilon_0\hbar^2}{\pi}$	$\frac{3\epsilon_0\hbar^2}{\pi}$	1				
f_{12}	$\frac{g_2}{g_1}\frac{2\pi\epsilon_0 m_e c^3}{e^2\omega_0^2}$	$\frac{2\epsilon_0 m_e\hbar\omega_0}{\pi e^2}$	$\frac{g_2}{g_1}\frac{2\epsilon_0 m_e\hbar\omega_0}{\pi e^2}$	$\frac{g_2}{g_1}\frac{2m_e\omega_0}{3e^2\hbar}$	1			
f_{21}	$-\frac{2\pi\epsilon_0 m_e c^3}{e^2\omega_0^2}$	$-\frac{g_1}{g_2}\frac{2\epsilon_0 m_e\hbar\omega_0}{\pi e^2}$	$-\frac{2\epsilon_0 m_e\hbar\omega_0}{\pi e^2}$	$-\frac{2m_e\omega_0}{3e^2\hbar}$	$-\frac{g_1}{g_2}$	1		
S_{12}	$g_2\frac{3\pi\epsilon_0\hbar c^3}{\omega_0^3}$	$g_1\frac{3\epsilon_0\hbar^2}{\pi}$	$g_2\frac{3\epsilon_0\hbar^2}{\pi}$	g_2	$g_1\frac{3\hbar e^2}{2m_e\omega_0}$	$-g_2\frac{3\hbar e^2}{2m_e\omega_0}$	1	
σ_{ω_0}	$\frac{g_2}{g_1}\frac{\pi^2 c^2}{\omega_0^2}$	$\frac{\hbar\omega_0}{\pi c}$	$\frac{g_2}{g_1}\frac{\hbar\omega_0}{\pi c}$	$\frac{g_2}{g_1}\frac{\omega_0}{3\epsilon_0\hbar c}$	$\frac{e^2}{2\epsilon_0 m_e c}$	$-\frac{g_2}{g_1}\frac{e^2}{2\epsilon_0 m_e c}$	$\frac{1}{g_1}\frac{\omega_0}{3\epsilon_0\hbar c}$	1

2.3 Further Reading

For calculation of the Einstein B coefficient from atomic properties, we have followed the development in

- R. Louden, *The Quantum Theory of Light,* 2nd edition, chapter 2, Clarendon Press, Oxford, 1983.

A comprehensive discussion of absorption and emission in real atoms, gas-phase laser action, and atomic spectroscopy with laser sources can be found in

- A. Corney, *Atomic and Laser Spectroscopy,* Clarendon Press, Oxford, 1977.

Useful older tables of line strengths and oscillator strengths for atoms from H to Ca can be found in

- *Atomic Transition Probabilities, Vols. I, II*, National Standard Reference Data Series, National Bureau of Standards (NSRDS-NBS 4,22), U.S. Government Printing Office, Washington, DC 20402. These tables have evolved into a continually updated database available on the National Institute of Standards and Technology (NIST) Website at *http://physics.nist.gov/PhysRefData/.*

A thorough discussion of the theory of absorption and emission of radiation from multilevel (real) atoms can be found in

- I. I. Sobelman, *Atomic Spectra and Radiative Transitions*, Springer-Verlag, Berlin, 1977.

Chapter 3

The Optical Bloch Equations

3.1 Introduction

So far we have concentrated on small-amplitude, broadband, phase-incoherent light fields interacting weakly with an atom or a collection of atoms in a dilute gas. Equations 2.19 and 2.20 provide formulas from which we can calculate the probability of finding a two-level atom in the excited state, but these expressions were developed by averaging over the spectral line width, ignoring any phase relation between the driving field and the driven dipole, and assuming essentially negligible depopulation of the ground state. For the first half of the twentieth century these assumptions corresponded to the light sources available in the laboratory, usually incandescent, arc, or plasma discharge lamps. After the invention of the laser in 1958, monomode and pulsed lasers quickly replaced lamps as the common source of optical excitation. These new light sources triggered an explosive revolution in optical science, the consequences of which continue to reverberate throughout physics, chemistry, electrical engineering, and biology. The characteristics of laser sources are far superior to those of the old lamps in every way. They are intense, highly directional, spectrally narrow, and phase-coherent. The laser has spawned a multitude of new spectroscopies, new disciplines such as quantum electronics, the study of the statistical properties of light in quantum optics, optical cooling and trapping of microscopic particles, control of chemical reactivity, and new techniques for ultra-high-resolution imaging and microscopies.

We are obliged therefore to examine what happens when our two-level atom interacts with these light sources, spectrally narrow compared to the natural width of optical transitions, with well-defined states of polarization and phase, and intensities sufficient to depopulate significantly the ground state. We seek an equation that will describe the time - evolution of well-defined two-level atoms interacting strongly with a single mode of the radiation field. Our initial thought might be to use the Schrödinger equation since it indeed describes the time-evolution of the state of any system defined. If we were interested only in stimulated processes, such as absorption of the single-mode wave incident on the atom, then the Schrödinger equation would suffice. The problem is that we want to describe relaxation as well as excitation processes, because in most realistic situations the atoms reach a steady state where the rate of excitation and relaxation equalize. Spontaneous emission (and any other dissipative process) therefore must be included in the physical description of the time-evolution of our light-plus-atom system. Now, however, we no longer have a system restricted to a single light-field mode (state) and two atom states. *Spontaneous emission populates a statistical distribution of light-field states and leaves the atom in a distribution of momentum states.* This situation cannot be described by a single wavefunction but only by some distribution of wavefunctions, and we can only hope to calculate the *probability* of finding the system among the distribution of state wavefunctions. The Schrödinger equation therefore no longer applies, and we have to seek the time-evolution of a system defined by a *density operator* which characterizes a statistical mixture of quantum states. The optical Bloch equations describe the time-evolution of the matrix elements of this density operator, and therefore we must use them in place of the Schrödinger equation. In order to appreciate the origin and physical content of the optical Bloch equations, we begin by reviewing the rudiments of density matrix theory.

3.2 The Density Matrix

3.2.1 Nomenclature and properties

We define a *density operator* ρ

$$\rho = \sum_i P_i \left|\psi_i\right\rangle \left\langle\psi_i\right| \tag{3.1}$$

where $\left|\psi_i\right\rangle$ is one of the complete set of orthonormal quantum states of some system, and we have a statistical distribution of these orthonormal states governed by the probability P_i of finding $\left|\psi_i\right\rangle$ in the state ensemble. The probability P_i, of course, lies between 0 and 1; and $\sum_i P_i = 1$. Note that the density operator acts on a member of the ensemble $\left|\psi_i\right\rangle$ to produce the probability of finding the system in $\left|\psi_i\right\rangle$,

$$\rho\left|\psi_i\right\rangle = \sum_i P_i \left|\psi_i\right\rangle \left\langle\psi_i|\psi_i\right\rangle = P_i \left|\psi_i\right\rangle \tag{3.2}$$

If all the members of the ensemble are in the same state, say, $\left|\psi_k\right\rangle$, then the density operator reduces to

$$\rho = \left|\psi_k\right\rangle \left\langle\psi_k\right|$$

and the system is said to be in a pure state with $P_k = 1$. From Eq. 3.2 we find the diagonal matrix elements of the density operator to be the probability of finding the system in state $|\psi_i\rangle$

$$\langle\psi_i|\,\rho\,|\psi_i\rangle = P_i$$

and, assuming all $|\psi_i\rangle$ to be orthonormal, the off-diagonal elements are necessarily zero. Furthermore

$$\sum_i \langle\psi_i|\,\rho\,|\psi_i\rangle = 1$$

3.2.2 Matrix representation

The next step is to develop *matrix representations* of the density operator by expanding the state vectors $|\psi_i\rangle$ in a complete orthonormal *basis set*

$$|\psi_i\rangle = \sum_n c_{ni}\,|n\rangle = \sum_n |n\rangle\,\langle n|\psi_i\rangle \tag{3.3}$$

where the closure relation is

$$\sum_n |n\rangle\,\langle n| = 1$$

and

$$\langle n|\psi_i\rangle = c_{ni}$$

is the projection of state vector $|\psi_i\rangle$ onto basis vector $|n\rangle$. Now we can write a matrix representation of the density operator in the basis $\{|n\rangle\}$ from the definition of ρ in Eq. 3.1 by substituting the basis set expansion of $|\psi_i\rangle$ and $\langle\psi_i|$ in Eq. 3.3:

$$\begin{aligned}\rho = \sum_i P_i\,|\psi_i\rangle\,\langle\psi_i| &= \sum_i P_i \sum_{nm} |n\rangle\,\langle n|\psi_i\rangle\,\langle\psi_i|m\rangle\,\langle m| \\ &= \sum_i P_i \sum_{nm} c_{ni}c^*_{mi}\,|n\rangle\,\langle m|\end{aligned} \tag{3.4}$$

The matrix elements of ρ in this representation are

$$\langle n|\rho|m\rangle = \sum_i P_i c_{ni} c^*_{mi} \tag{3.5}$$

with the diagonal matrix elements

$$\langle n|\rho|n\rangle = \sum_i P_i\,|c_{ni}|^2 \tag{3.6}$$

and

$$\langle n|\rho|m\rangle^* = \sum_i P_i c^*_{ni} c_{mi} = \sum_i P_i \sum_{mn} \langle m|\psi_i\rangle\,\langle\psi_i|n\rangle = \langle m|\rho|n\rangle$$

which means that the ρ operator is Hermitian. For a simple system such as our two-level atom that, without spontaneous emission, can be described by a single wavefunction, Eqs. 3.4, 3.5, and 3.6, respectively reduce to

$$\rho = \sum_{nm} c_{ni} c_{mi}^* |n\rangle \langle m| \tag{3.7}$$

$$\langle n|\rho|m\rangle = c_{ni} c_{mi}^* \tag{3.8}$$

$$\langle n| \rho |n\rangle = |c_{ni}|^2 \tag{3.9}$$

The sum of the diagonal elements of the representation matrix is called the *trace,* and it is a fundamental property of the density operator because it is invariant to any unitary transformation of the representation:

$$\mathrm{Tr}\ \rho \equiv \sum_n \langle n| \rho |n\rangle \tag{3.10}$$

From the definition of the density operator, Eq. 3.1, we can write Eq. 3.10 as

$$\mathrm{Tr}\ \rho = \sum_{ni} P_i \langle n|\psi_i\rangle \langle \psi_i|n\rangle$$

Then reversing the two matrix element factors and using the closure relation

$$\mathrm{Tr}\ \rho = \sum_{ni} P_i \langle \psi_i|n\rangle \langle n|\psi_i\rangle = \sum_i P_i \langle \psi_i|\psi_i\rangle = 1$$

which shows that the trace of the representation of the density operator is equal to unity, independent of the basis for the matrix representation.

The ensemble averages of observables are expressed as

$$\langle \hat{O}\rangle = \sum_i P_i \langle \psi_i| \hat{O} |\psi_i\rangle$$

but

$$\rho\hat{O} = \sum_i P_i |\psi_i\rangle \langle \psi_i| \hat{O}$$

and in the basis $\{|n\rangle\}$

$$\langle n| \rho\hat{O} |m\rangle = \langle n| \sum_i P_i |\psi_i\rangle \langle \psi_i| \hat{O} |m\rangle$$
$$= \sum_i P_i \langle n|\psi_i\rangle \langle \psi_i| \hat{O} |m\rangle = \sum_i P_i \langle \psi_i|n\rangle \langle m| \hat{O} |\psi_i\rangle$$

where have assumed that the operator of the physical observable $\hat{O}$ is Hermitian and that the representation of the product of two Hermitian operators $\rho\hat{O}$ is Hermitian. Now, along the diagonal we have

$$\langle n| \rho\hat{O} |n\rangle = \sum_i P_i \langle \psi_i|n\rangle \langle n| \hat{O} |\psi_i\rangle$$

With the closure condition on the basis set $\{|n\rangle\}$, we then have

$$\text{Tr } \langle n| \rho\hat{O} |n\rangle = \sum_i P_i \langle\psi_i| \hat{O} |\psi_i\rangle = \langle\hat{O}\rangle \tag{3.11}$$

Equation 3.11 says that an ensemble average of any dynamical observable $\hat{O}$ can be calculated from the on-diagonal matrix elements of the operator $\rho\hat{O}$. Since the trace is independent of the basis, any unitary transformation that carries the matrix representation from basis $\{|n\rangle\}$ to some other basis $\{|t\rangle\}$ leaves the trace invariant. Using the definition of a unitary transformation, one can easily show that the trace of a cyclic permutation of a product of operators is invariant. For example

$$\text{Tr } [ABC] = \text{Tr } [CAB] = \text{Tr } [BAC]$$

and in particular

$$\text{Tr } \left[\rho\hat{O}\right] = \text{Tr } \left[\hat{O}\rho\right] = \left\langle\hat{O}\right\rangle$$

3.2.3 Review of operator representations

We will see that the optical Bloch equations (Eqs. 4.50–4.53) are a set of coupled differential equations relating the time dependence of different matrix elements of a density operator. It seems worthwhile, therefore, to review commonly encountered "representations" of the time dependence of operators, quantum states, and ensembles of quantum states. The optical Bloch equations present somewhat different forms depending on the representation in which they are expressed.

The *Schrödinger representation* of the time evolution of a quantum system is expressed by the familiar Schrödinger equation

$$\hat{H} |\psi(\mathbf{r},t)\rangle = i\hbar\frac{\partial}{\partial t} |\psi(\mathbf{r},t)\rangle \tag{3.12}$$

in which all the time dependence resides in the state functions, and the operators that stand for the dynamical variables (energy, angular momentum, position, etc.) are independent of time. In the *Heisenberg representation* all the explicit time dependence resides in the operators and the state functions are time-independent. The *interaction representation* is a hybrid of the Schrödinger and Heisenberg representations appropriate for Hamiltonians of the form

$$\hat{H} = \hat{H}_0 + \hat{V}(t)$$

where $\hat{H}_0$ is a time-independent Hamiltonian of the unperturbed system and $\hat{V}(t)$ is a time-dependent coupling interaction, often a perturbing oscillatory field.

Time evolution operator

Recall from elementary quantum mechanics the time evolution operator, $U(t, t_0)$, which acts on the ket space of a quantum state to transform it from initial time t_0 to a later time t:.

$$U(t, t_0) \left|\psi(\mathbf{r},t_0)\right\rangle = \left|\psi(\mathbf{r},t)\right\rangle \tag{3.13}$$

Here are a few properties of the time - evolution operator. Note first that

$$U(t_2, t_0) = U(t_2, t_1)U(t_1, t_0)$$

where $t_0 < t_1 < t_2$. Note also that the time - reversal operation

$$\left|\psi(\mathbf{r},t_0)\right\rangle = U(t_0, t) \left|\psi(\mathbf{r},t)\right\rangle$$

together with multiplication from the left by $U^{-1}(t, t_0)$ of Eq. 3.13 implies

$$U(t_0, t) = U^{-1}(t, t_0)$$

The conjugate time - evolution operator acts on the bra space:

$$\left\langle\psi(\mathbf{r},t)\right| = \left\langle\psi(\mathbf{r},t_0)\right| U^{\dagger}(t, t_0) \tag{3.14}$$

If the Hamiltonian is time - independent, then we can see from a formal integration of Eq. 3.12 that

$$U(t, t_0) = e^{-i\widehat{H}(t-t_0)/\hbar} \tag{3.15}$$

and

$$U^{\dagger}(t, t_0) = e^{i\widehat{H}(t-t_0)/\hbar} \tag{3.16}$$

so that from Eqs. 3.13 and 3.14

$$\left|\psi(\mathbf{r},t)\right\rangle = e^{-i\widehat{H}(t-t_0)/\hbar} \left|\psi(\mathbf{r},t_0)\right\rangle \tag{3.17}$$

and

$$\left\langle\psi(\mathbf{r},t)\right| = \left\langle\psi(\mathbf{r},t_0)\right| e^{i\widehat{H}(t-t_0)/\hbar} \tag{3.18}$$

From Eqs. 3.15 and 3.16 $U^{\dagger}U$ is a time-independent constant that we set equal to unity for normalization

$$U^{\dagger}(t, t_0)U(t, t_0) = 1$$

from which we obtain the unitarity property by multiplication of $U^{-1}(t, t_0)$ from the right:

$$U^{\dagger}(t, t_0) = U^{-1}(t, t_0) \tag{3.19}$$

The unitarity property is important because it can be used in similarity transformations to change the representation of operators from one basis to another. If $\left|\psi_i(\mathbf{r},t)\right\rangle$ is an eigenstate of the Hamiltonian, then it can be shown that

$$\begin{aligned} U(t, t_0) \left|\psi_i(\mathbf{r},t)\right\rangle &= e^{-i\widehat{H}(t-t_0)/\hbar} \left|\psi_i(\mathbf{r},t)\right\rangle \\ &= e^{-iE_i(t-t_0)/\hbar} \left|\psi_i(\mathbf{r},t)\right\rangle = e^{-i\omega(t-t_0)} \left|\psi_i(\mathbf{r},t)\right\rangle \end{aligned}$$

and similarly for $U^{\dagger}(t, t_0)$ operating in the bra space.

Heisenberg representation

We express the Heisenberg representation of operators and quantum states through a unitary transformation of the Schrödinger representation. Starting from

$$|\varphi(\mathbf{r})\rangle = U^{\dagger}(t, t_0)\,|\psi(\mathbf{r},t)\rangle \tag{3.20}$$

we examine the time-dependence of $|\varphi(\mathbf{r})\rangle$ by differentiating both sides of Eq. 3.20:

$$\frac{\partial\,|\varphi(\mathbf{r})\rangle}{\partial t} = \frac{dU^{\dagger}(t,t_0)}{dt}\,|\psi(\mathbf{r},t)\rangle + U^{\dagger}(t,t_0)\frac{\partial\,|\psi(\mathbf{r},t)\rangle}{\partial t} \tag{3.21}$$

Now from the definitions of $\hat{H}$ and U, (Eqs. 3.12, 3.13)

$$\frac{dU}{dt} = \frac{1}{i\hbar}\hat{H}U \tag{3.22}$$

and

$$\frac{dU^{\dagger}}{dt} = -\frac{1}{i\hbar}U^{\dagger}\hat{H} \tag{3.23}$$

where we assume that the Hamiltonian operator is Hermitian, $\hat{H} = \hat{H}^{\dagger}$. Substituting $\frac{dU^{\dagger}}{dt}$ from Eq. 3.23 and $\frac{\partial|\psi(\mathbf{r},t)\rangle}{\partial t}$ from the Schrödinger equation (Eq. 3.12) into 3.21, we see that

$$\frac{\partial\,|\varphi(\mathbf{r})\rangle}{\partial t} = 0$$

or, in other words, the operation of $U^{\dagger}(t,t_0)$ on $|\psi(\mathbf{r},t)\rangle$ removes any time dependence of the wavefunction $|\varphi(\mathbf{r})\rangle$. By unitarity, and from Eq. 3.20 we also have

$$U(t,t_0)\,|\varphi(\mathbf{r})\rangle = |\psi(\mathbf{r},t)\rangle$$

Now we can write the matrix element for any operator $\hat{O}$

$$\langle\psi(\mathbf{r},t)|\,\hat{O}\,|\psi(\mathbf{r},t)\rangle = \langle\varphi(\mathbf{r})|\,U^{\dagger}(t,t_0)\hat{O}U(t,t_0)\,|\varphi(\mathbf{r})\rangle$$

We see that the matrix element of the operator $\hat{O}$ in the Schrödinger representation with time-dependent basis $\{|\psi(\mathbf{r},t)\rangle\}$ is equal to the matrix element of the operator $U^{\dagger}(t,t_0)\hat{O}U(t,t_0)$ in the Heisenberg representation with time-independent basis $\{|\varphi(\mathbf{r})\rangle\}$. More succinctly, we can write

$$\hat{O}_{HR} = U^{\dagger}(t,t_0)\hat{O}_{SR}U(t,t_0) \tag{3.24}$$

and

$$U(t,t_0)\hat{O}_{HR}U^{\dagger}(t,t_0) = \hat{O}_{SR} \tag{3.25}$$

where the subscripts HR and SR mean "Heisenberg representation" and "Schrödinger representation," respectively.

Just as the Schrödinger equation expresses the time evolution of a quantum state operated on by the Hamiltonian in the Schrödinger representation, the

Heisenberg equation expresses the time evolution of an operator in the Heisenberg representation acting on time-independent quantum states:

$$i\hbar\frac{d\hat{O}_{HR}}{dt} = i\hbar\frac{d}{dt}\left[U^{\dagger}(t,t_0)\hat{O}_{SR}U(t,t_0)\right]$$
$$= i\hbar\left[\frac{dU^{\dagger}}{dt}\hat{O}_{SR}U(t,t_0) + U^{\dagger}(t,t_0)\hat{O}_{SR}\frac{dU}{dt}\right]$$

Substituting from Eqs. 3.22 , 3.23 , and 3.25, we find

$$i\hbar\frac{d\hat{O}_{HR}}{dt} = -\hat{H}U(t,t_0)\hat{O}_{HR}U^{\dagger}(t,t_0) + U(t,t_0)\hat{O}_{HR}U^{\dagger}\hat{H} \quad (3.26)$$
$$\frac{d\hat{O}_{HR}}{dt} = \frac{i}{\hbar}\left[\hat{H},\hat{O}_{HR}\right] \quad (3.27)$$

Thus the time rate of change of an operator in the Heisenberg representation is given by the commutator of that operator with the total Hamiltonian of the system. Note that if an operator representing a dynamical variable commutes with the Hamiltonian in the Schrödinger representation, it will also commute with the Hamiltonian in the Heisenberg representation, and therefore for the complete set of commuting observables, we obtain

$$i\hbar\frac{d\hat{O}_{HR}}{dt} = \left[\hat{O}_{HR},\hat{H}\right] = 0$$

From Eqs. 3.15 and 3.16 we can also write

$$\hat{O}_{HR} = e^{i\hat{H}(t-t_0)/\hbar}\hat{O}_{SR}e^{-i\hat{H}(t-t_0)/\hbar} \quad (3.28)$$

or

$$e^{-i\hat{H}(t-t_0)/\hbar}\hat{O}_{HR}e^{i\hat{H}(t-t_0)/\hbar} = \hat{O}_{SR}$$

and if $\{|\psi(\mathbf{r},t)\rangle\}$ and $\{|\varphi(\mathbf{r})\rangle\}$ are bases of eigenstates of $\hat{H}$, we have for any two states n, m

$$\langle\psi_n(\mathbf{r},t_0)|\, e^{-i\hat{H}(t-t_0)/\hbar}\hat{O}_{SR}e^{i\hat{H}(t-t_0)/\hbar}\,|\psi_m(\mathbf{r},t_0)\rangle = \quad (3.29)$$
$$e^{-i/\hbar(E_n-E_m)t}\,\langle\psi_n(\mathbf{r},t)|\,\hat{O}_{SR}\,|\psi_m(\mathbf{r},t)\rangle = \quad (3.30)$$
$$\langle\varphi_n(\mathbf{r})|\,\hat{O}_{HR}\,|\varphi_m(\mathbf{r})\rangle \quad (3.31)$$

Thus the matrix elements of an operator in the Schrödinger and Heisenberg representations are related by a simple phase factor.

Interaction representation The interaction representation treats problems where the total Hamiltonian is composed of a time-independent part and a time-dependent term:

$$\hat{H} = \hat{H}_0 + \hat{V}(t)$$

Analogous to Eq. 3.17, we define a time-evolution operator in terms of the *time-independent part* of the total Hamiltonian:

$$\left|\tilde{\psi}(\mathbf{r},t)\right\rangle = e^{i\hat{H}_0(t-t_0)/\hbar}\left|\psi(\mathbf{r},t)\right\rangle \tag{3.32}$$

$$\tilde{O}(t) = e^{i\hat{H}_0(t-t_0)/\hbar}\hat{O}e^{-i\hat{H}_0(t-t_0)/\hbar} \tag{3.33}$$

Now we seek the time-dependence of quantum states and operators in the interaction representation. From Eq. 3.32 we can get the inverse relation

$$\left|\psi(\mathbf{r},t)\right\rangle = e^{-i\hat{H}_0(t-t_0)/\hbar}\left|\tilde{\psi}(\mathbf{r},t)\right\rangle$$

and substitution into the Schrödinger equation (Eq. 3.12) yields

$$\frac{\partial}{\partial t}\left|\tilde{\psi}(\mathbf{r},t)\right\rangle = -i\frac{\hat{V}(t)}{\hbar}\left|\tilde{\psi}(\mathbf{r},t)\right\rangle$$

We see that in the interaction representation only the perturbation term of the Hamiltonian controls the time evolution. Taking the time derivative of both sides of the defining equation for the operator $\tilde{O}$ in the interaction representation (Eq. 3.33) results in

$$\frac{d\tilde{O}}{dt} = \frac{i}{\hbar}\left[\hat{H}_0, \tilde{O}\right]$$

So we see that the time derivative can be expressed in the form of a commutator, similar to the Heisenberg equation (Eq. 3.27) except that only the unperturbed term of the Hamiltonian is in the argument of the commutator operator. It is also clear that, similar to Eq. 3.28, we have

$$\hat{O}_{IR} = e^{i\hat{H}_0(t-t_0)/\hbar}\hat{O}_{SR}e^{-i\hat{H}_0(t-t_0)/\hbar}$$

Note that the transformation between the interaction and Schrödinger representations involves only $\hat{H}_0$ in the exponential factors and not $\hat{H}$. It is also clear that the off-diagonal matrix elements between the two representations are related by a simple phase factor

$$e^{-i/\hbar(E_n-E_m)t}\left\langle\psi_n(\mathbf{r},t)\right|\hat{O}_{SR}\left|\psi_m(\mathbf{r},t)\right\rangle = \left\langle\varphi_n(\mathbf{r})\right|\hat{O}_{IR}\left|\varphi_m(\mathbf{r})\right\rangle \tag{3.34}$$

where the eigenvalues $E_{n,m}$ in the exponential factors, $e^{-i/\hbar(E_n-E_m)t}$ are the energies of the unperturbed Hamiltonian H_0.

3.2.4 Time dependence of the density operator

Going back to the definition of the density operator (Eq. 3.1), we can express its time dependence in terms of time-dependent quantum states and the time-evolution operator,

$$\begin{aligned}\rho(t) &= \sum_i P_i\left|\psi_i(t)\right\rangle\left\langle\psi_i(t)\right| \qquad (3.35)\\ &= \sum_i P_i U(t,t_0)\left|\psi_i(t_0)\right\rangle\left\langle\psi_i(t_0)\right|U^\dagger(t,t_0)\end{aligned}$$

and writing

$$\rho(t_0) = \sum_i P_i \left|\psi_i(t_0)\right\rangle \left\langle\psi_i(t_0)\right|$$

we see immediately that

$$\rho(t) = U(t, t_0)\rho(t_0)U^\dagger(t, t_0) \tag{3.36}$$

and for the common case of a time-independent Hamiltonian:

$$\rho(t) = e^{-i\hat{H}(t-t_0)/\hbar}\rho(t_0)e^{i\hat{H}(t-t_0)/\hbar}$$

Now we find the time derivative of the density operator by differentiating both sides of Eq. 3.36 and substituting Eqs. 3.23 and 3.22 for the time derivatives of U and $U^\dagger$. The result is

$$\frac{d\rho(t)}{dt} = \frac{i}{\hbar}\left[\rho(t), \hat{H}\right] \tag{3.37}$$

The commutator itself can be considered an operator, so we can write

$$\hat{L}\rho(t) = \frac{i}{\hbar}\left[\rho(t), \hat{H}\right] \tag{3.38}$$

where $\hat{L}$ is called the *Liouville operator* and Eq. 3.37 is called the *Liouville equation. The Liouville equation describes the time evolution of the density operator, which itself specifies the distribution of an ensemble of quantum states subject to the Hamiltonian operator* $\hat{H}$. Although the Liouville equation resembles the Heisenberg equation in form, Eq. 3.35 shows that $\rho(t)$ is in the Schrödinger representation.

Now we can transform the density operator to the interaction representation

$$\tilde{\rho}(t)_{IR} = e^{i\hat{H}_0(t-t_0)/\hbar}\,\rho(t)_{SR}\,e^{-i\hat{H}_0(t-t_0)/\hbar} \tag{3.39}$$

and seek the time rate of change of $\tilde{\rho}(t)$ analogous to the Liouville equation. Taking the time derivative of both sides of Eq. 3.39 and substituting Eq. 3.27 for $\frac{d\rho}{dt}$ results in

$$\frac{d\tilde{\rho}(t)}{dt} = \frac{i}{\hbar}\left[\tilde{\rho}(t), \hat{V}(t)\right] \tag{3.40}$$

Equation 3.40 shows that the time evolution of the density operator in the interaction representation depends only on the time-dependent part of the total Hamiltonian. For a two-level atom interacting perturbatively with a light field, the Hamiltonian is

$$\hat{H} = \hat{H}_A + \hat{V}(t) = \hat{H}_A + \hat{\boldsymbol{\mu}} \cdot \mathbf{E_0}\cos\omega t$$

where $\hat{H}_A$ is the atomic structure part of the Hamiltonian and $\hat{V}(t)$ is the transition dipole interaction with the classical oscillating electric field. The interaction representation is the natural choice for this type of problem.

3.2.5 Density operator matrix elements

Since the optical Bloch equations are coupled differential equations relating the matrix elements of the density operator, we need to examine the time dependence of these matrix elements, based on what we have established for the density operator itself. We start with the Liouville equation (Eq. 3.37) and take matrix elements of this operator equation

$$\langle m| \frac{d\rho(t)}{dt} |n\rangle = \frac{i}{\hbar} \langle m| \left[\rho(t), \hat{H}\right] |n\rangle = \frac{i}{\hbar} \langle m| \left[\rho(t), \hat{H}_A + \hat{V}\right] |n\rangle \tag{3.41}$$
$$= \frac{i}{\hbar} (E_n - E_m) \langle m| \rho(t) |n\rangle + \frac{i}{\hbar} \langle m| \left[\rho(t), \hat{V}\right] |n\rangle$$

where $|m\rangle$ and $|n\rangle$ are members of a complete set of basis vectors $\{|k\rangle\}$ that are also eigenkets of $\widehat{H}_A$ and span the space of $\hat{H}$. Now we insert the closure relation $\sum_k |k\rangle \langle k|$ into the commutator on the left side of Eq. 3.41:

$$\langle m| \left[\rho(t), \hat{V}\right] |n\rangle = \sum_k \langle m| \rho(t) |k\rangle \langle k| \hat{V} |n\rangle - \langle m| \hat{V} |k\rangle \langle k| \rho(t) |n\rangle \tag{3.42}$$

For our two-level atom the complete set includes only two states: $|1(t)\rangle = |1\rangle$ and $|2(t)\rangle = e^{-i\omega_0 t} |2\rangle$. Furthermore the matrix elements of the dipole coupling operator $\hat{V}$ are only off-diagonal, $\langle 1| \hat{V} |2\rangle$ and $\langle 2| \hat{V} |1\rangle$ with $\hat{V}$ Hermitian: $\left(\langle 1| \hat{V} |2\rangle\right)^* = \langle 2| \hat{V}^* |1\rangle$. The commutator matrix elements in Eq. 3.42 simplify to

$$\langle 1| \left[\rho(t), \hat{V}\right] |1\rangle = \langle 1| \rho |2\rangle \langle 2| \hat{V} |1\rangle - \langle 1| \hat{V} |2\rangle \langle 2| \rho |1\rangle$$
$$= \rho_{12} V_{21} - V_{12} \rho_{21}$$

Similarly

$$\langle 2| \left[\rho(t), \hat{V}\right] |2\rangle = \langle 2| \rho |1\rangle \langle 1| \hat{V} |2\rangle - \langle 2| \hat{V} |1\rangle \langle 1| \rho |2\rangle$$
$$= \rho_{21} V_{12} - V_{21} \rho_{12}$$

for the off-diagonal matrix elements

$$\langle 1| \left[\rho(t), \hat{V}\right] |2\rangle = \langle 1| \rho |1\rangle \langle 1| \hat{V} |2\rangle - \langle 1| \hat{V} |2\rangle \langle 2| \rho |2\rangle$$
$$= V_{12} (\rho_{11} - \rho_{22})$$

and

$$\langle 2| [\rho(t), V] |1\rangle = \langle 2| \rho |2\rangle \langle 2| V |1\rangle - \langle 2| V |1\rangle \langle 1| \rho |1\rangle$$
$$= V_{21} (\rho_{22} - \rho_{11})$$

so that Eq. 3.41 takes the form

$$\frac{d\rho_{11}}{dt} = \frac{i}{\hbar}\left[\rho_{12}V_{21} - V_{12}\rho_{21}\right] \tag{3.43}$$
$$\frac{d\rho_{22}}{dt} = \frac{i}{\hbar}\left[\rho_{21}V_{12} - V_{21}\rho_{12}\right]$$
$$\frac{d\rho_{12}}{dt} = i\omega_0\rho_{12} + \frac{i}{\hbar}\left[V_{12}\left(\rho_{11} - \rho_{22}\right)\right] \tag{3.44}$$
$$\frac{d\rho_{21}}{dt} = -i\omega_0\rho_{21} + \frac{i}{\hbar}\left[V_{21}\left(\rho_{22} - \rho_{11}\right)\right]$$

and we see that

$$\frac{d\rho_{12}^*}{dt} = \frac{d\rho_{21}}{dt}$$

The set of equations 3.43 constitute the optical Bloch equations in the Schrödinger representation. They do not include loss terms from spontaneous emission. We transform the optical Bloch equations to the interaction representation by replacing the Liouville equation (Eq. 3.37) with Eq. 3.40 and taking the matrix elements:

$$\frac{d\tilde{\rho}_{11}}{dt} = \frac{i}{\hbar}\left[\tilde{\rho}_{12}V_{21} - V_{12}\tilde{\rho}_{21}\right] \tag{3.45}$$
$$\frac{d\tilde{\rho}_{22}}{dt} = \frac{i}{\hbar}\left[\tilde{\rho}_{21}V_{12} - V_{21}\tilde{\rho}_{12}\right] \tag{3.46}$$
$$\frac{d\tilde{\rho}_{12}}{dt} = \frac{i}{\hbar}\left[V_{12}\left(\tilde{\rho}_{11} - \tilde{\rho}_{22}\right)\right]$$
$$\frac{d\tilde{\rho}_{21}}{dt} = \frac{i}{\hbar}\left[V_{21}\left(\tilde{\rho}_{22} - \tilde{\rho}_{11}\right)\right]$$

The interaction representation simplifies the expressions for the time dependence of the coherences by eliminating the first term on the right side. Transforming to the interaction representation removes the time dependence of the basis vectors spanning the space of our two-level atom.

We have established the optical Bloch equations from the Liouville equation, the fundamental equation of motion of the density operator, and we have seen how a unitary transformation can be used to "represent" these equations in either the Schrödinger, Heisenberg, or interaction representations. So far the optical Bloch equations do not include the possibility of spontaneous emission. We will discuss how to include this effect in Section 4.5.

We show in Section 4 how we can supplement this somewhat formal development by constructing the optical Bloch equations for a two-level system, starting from the expansion coefficients of our two-level wavefunction, Eq. 2.3.

3.2.6 Time evolution of the density matrix

The equation of motion of the density matrix is given by the *Liouville equation,*as discussed in Section 3.2.4 (Eq. 3.37),

$$\frac{d\rho(t)}{dt} = \frac{i}{\hbar}\left[\rho(t), \widehat{H}\right] \tag{3.47}$$

and the time - dependence of the matrix elements for any two-level system subject to some off-diagonal coupling V_{12} between the ground and excited levels separated by an energy $\hbar\omega_0$ is given by

$$\frac{d\rho_{11}}{dt} = \frac{i}{\hbar}\left[\rho_{12}V_{21} - V_{12}\rho_{21}\right] \tag{3.48}$$

$$\frac{d\rho_{12}}{dt} = i\omega_0\rho_{12} + \frac{i}{\hbar}\left[V_{12}\left(\rho_{11} - \rho_{22}\right)\right]$$

$$\frac{d\rho_{21}}{dt} = -i\omega_0\rho_{21} + \frac{i}{\hbar}\left[V_{21}\left(\rho_{22} - \rho_{11}\right)\right]$$

$$\frac{d\rho_{22}}{dt} = \frac{i}{\hbar}\left[\rho_{21}V_{12} - V_{21}\rho_{12}\right] = -\frac{d\rho_{11}}{dt} \tag{3.49}$$

These equations can be written in matrix form,

$$\begin{bmatrix} \frac{d\rho_{11}}{dt} \\ \frac{d\rho_{12}}{dt} \\ \frac{d\rho_{21}}{dt} \\ \frac{d\rho_{22}}{dt} \end{bmatrix} = \frac{i}{\hbar}\begin{bmatrix} 0 & V_{21} & -V_{12} & 0 \\ V_{12} & \hbar\omega_0 & 0 & -V_{12} \\ -V_{21} & 0 & -\hbar\omega_0 & V_{21} \\ 0 & -V_{21} & +V_{12} & 0 \end{bmatrix}\begin{bmatrix} \rho_{11} \\ \rho_{12} \\ \rho_{21} \\ \rho_{22} \end{bmatrix} \tag{3.50}$$

or as a vector cross-product

$$\frac{d\boldsymbol{\beta}}{dt} = -\boldsymbol{\beta} \times \boldsymbol{\Omega} \tag{3.51}$$

where

$$\boldsymbol{\beta} = \hat{\boldsymbol{\imath}}\left(\rho_{21} + \rho_{12}\right) + \hat{\boldsymbol{\jmath}}i\left(\rho_{21} - \rho_{12}\right) + \hat{\mathbf{k}}\left(\rho_{22} - \rho_{11}\right) \tag{3.52}$$

and

$$\boldsymbol{\Omega} = \frac{1}{\hbar}\left[\hat{\boldsymbol{\imath}}\left(V_{21} + V_{12}\right) + \hat{\boldsymbol{\jmath}}i\left(V_{21} - V_{12}\right) + \hat{\mathbf{k}}\hbar\omega_0\right] \tag{3.53}$$

so that

$$\begin{aligned} \frac{d\boldsymbol{\beta}}{dt} &= \hat{\boldsymbol{\imath}}\frac{1}{\hbar}\left[i\hbar\omega_0\left(\rho_{12} - \rho_{21}\right) + \left(V_{21} - V_{12}\right)\left(\rho_{22} - \rho_{11}\right)\right] \\ &+ \hat{\boldsymbol{\jmath}}\frac{1}{\hbar}\left[\hbar\omega_0\left(\rho_{21} + \rho_{12}\right) - \left(V_{21} + V_{12}\right)\left(\rho_{22} - \rho_{11}\right)\right] \\ &+ \hat{\mathbf{k}}\frac{2}{\hbar}\left[i\left(V_{21}\rho_{12} - V_{12}\rho_{21}\right)\right] \end{aligned}$$

The vector $\boldsymbol{\beta}$ is called the *Bloch vector*, and its Cartesian components are often expressed as

$$\begin{aligned} u_1 &= \rho_{21} + \rho_{12} \\ u_2 &= i\left(\rho_{12} - \rho_{21}\right) \\ u_3 &= \rho_{11} - \rho_{22} \end{aligned} \tag{3.54}$$

In the case of a real coupling operator $V_{12} = V_{21}^{*}$, and the explicit equations of motion for the Bloch vector components become

$$\begin{aligned} \frac{d\beta_x}{dt} &= -\omega_0 u_2 \\ \frac{d\beta_y}{dt} &= -\omega_0 u_1 + \frac{2}{\hbar} V_{12} u_3 \\ \frac{d\beta_z}{dt} &= -\frac{2}{\hbar} V_{12} u_2 \end{aligned} \tag{3.55}$$

We have introduced the Bloch vector here to complete the formal presentation of the density matrix theory. The physical content and the usefulness of the Bloch vector will become clearer when we use this formalism to analyze electric and magnetic dipole couplings.

3.3 Further Reading

There are many excellent presentations of density matrix theory. For optical and collisional interactions, two quite useful books are

- M. Weissbluth, *Photon-Atom Interactions*, Academic Press, Boston, 1989.
- K. Blum, *Density Matrix Theory and Applications*, Plenum Press, New York, 1981.

Chapter 4

Optical Bloch Equations of a Two-Level Atom

4.1 Introduction

In this chapter we will begin to apply the ideas and tools we have established in Chapters 1–3. We will first apply the density matrix to a two-level atom coupled to a single-mode field without spontaneous emission. We will then introduce the atom Bloch vector as convenient and easily visualized way to describe the time–evolution of the coupled two-level atom. Next we introduce spontaneous emission; and, with Sections 4.4 and 4.4.1, introduce the important idea of polarization and susceptibility as the result of a collection of driven oscillating dipoles. The OBEs including spontaneous emission are then written down, and their steady-state solutions discussed. Dissipative processes always broaden transition lines, and we will discuss various broadening mechanisms in the last section.

4.2 Coupled differential equations

Now that we have established the language of density matrix theory, let us consider first the density matrix of our two-level atom in a pure state (and without spontaneous emission) in the $\{\Psi_1, \Psi_2\}$ representation. We recall Eq. 2.3, the time-dependent wavefunction of our two-level system

$$\begin{aligned}\Psi(\mathbf{r},t) &= C_1(t)\Psi_1(\mathbf{r},t) + C_2(t)\Psi_2(\mathbf{r},t) &\quad (4.1)\\ &= C_1(t)\psi_1(\mathbf{r})e^{-i\omega_1 t} + C_2(t)\psi_2(\mathbf{r})e^{-i\omega_2 t} &\quad (4.2)\end{aligned}$$

and Eqs. 2.10 and 2.11 describing the optical coupling:

$$\hbar\Omega_0 \cos\omega t\, e^{-i\omega_0 t} C_2 = i\frac{dC_1}{dt}$$
$$\hbar\Omega_0^* \cos\omega t e^{i\omega_0 t} C_1 = i\frac{dC_2}{dt}$$

We take the time-dependent form of the quantum state, $\Psi_n(\mathbf{r},t) = \psi_n(\mathbf{r})e^{-i\omega_n t}$ and write $\Psi_1(\mathbf{r},t)$ and $\Psi_2(\mathbf{r},t)$ as the basis states for the representation of the density operator ρ,. Then, following Eq. 3.7, we write

$$\begin{aligned} \rho_{11} &= |C_1|^2 \\ \rho_{22} &= |C_2|^2 \\ \rho_{12} &= C_1 C_2^* \\ \rho_{21} &= C_2 C_1^* \end{aligned} \tag{4.3}$$

which we form into a density matrix as

$$\begin{bmatrix} \rho_{11} & \rho_{12} \\ \rho_{21} & \rho_{22} \end{bmatrix}$$

Remembering the interpretation $|C_n|^2$ as the probability density of finding the atom in level n, the trace (sum of the diagonal elements) is equal to unity,

$$\rho_{11} + \rho_{22} = 1$$

These diagonal terms are called *populations*. We also have

$$\rho_{21} = \rho_{12}^*$$

The off-diagonal terms are called *coherences*.

Now we differentiate Eqs. 4.3 on both sides with respect to time

$$\begin{aligned} \frac{d\rho_{11}}{dt} &= C_1\frac{dC_1^*}{dt} + \frac{dC_1}{dt}C_1^* \\ \frac{d\rho_{22}}{dt} &= C_2\frac{dC_2^*}{dt} + \frac{dC_2}{dt}C_2^* \\ \frac{d\rho_{12}}{dt} &= C_1\frac{dC_2^*}{dt} + \frac{dC_1}{dt}C_2^* \\ \frac{d\rho_{21}}{dt} &= C_2\frac{dC_1^*}{dt} + \frac{dC_2}{dt}C_1^* \end{aligned} \tag{4.4}$$

and if we substitute Eqs. and 2.10, 2.11 for $\frac{dC_1}{dt}$ and $\frac{dC_2}{dt}$, , make the rotating-wave approximation, set $\Omega_0^* = \Omega_0$ and define the detuning $\Delta\omega \equiv \omega - \omega_0$, we find

$$\begin{aligned} \frac{d\rho_{22}}{dt} &= i\frac{\Omega_0}{2}\left[e^{i\Delta\omega t}\rho_{21} - e^{-i\Delta\omega t}\rho_{12}\right] = -\frac{d\rho_{11}}{dt} \\ \frac{d\rho_{12}}{dt} &= i\frac{\Omega_0}{2}e^{i\Delta\omega t}\left(\rho_{11} - \rho_{22}\right) = \frac{d\rho_{21}^*}{dt} \end{aligned} \tag{4.5}$$

Equations 4.5 describe the time-evolution of the on-diagonal and off-diagonal density matrix elements and constitute our first expressions for the optical Bloch equations not including spontaneous emission. For arbitrary initial conditions the solutions for ρ_{22} and ρ_{12} are not simple, but if we start with a collection of atoms in the ground state with the coupling light turned off, then the initial conditions are

$$\rho_{11} = 1 \qquad \rho_{22} = 0 \qquad \rho_{12} = 0$$

and the solution for the final excited-state population is

$$\rho_{22} = 1 - \rho_{11} = \frac{\Omega_0^2}{\Omega^2} \sin^2 \left(\frac{\Omega}{2} t \right) \tag{4.6}$$

and

$$\rho_{12} = \rho_{21}^* = e^{i\Delta\omega t} \frac{\Omega_0}{\Omega^2} \sin \left(\frac{\Omega}{2} t \right) \left[\Delta\omega \sin \left(\frac{\Omega}{2} t \right) + i\Omega \cos \left(\frac{\Omega}{2} t \right) \right] \tag{4.7}$$

with

$$\Omega = \sqrt{(\Delta\omega)^2 + \Omega_0^2} \tag{4.8}$$

where Ω is called the *Rabi frequency*. For the special (but frequent) case of on-resonance excitation $\omega = \omega_0$

$$\rho_{22} = \sin^2 \frac{\Omega_0}{2} t = \frac{1}{2}(1 - \cos \Omega_0 t) \tag{4.9}$$

with the *on-resonance Rabi frequency*

$$\Omega = \Omega_0 \tag{4.10}$$

while the on-resonance coherence becomes

$$\rho_{12} = i \sin \left(\frac{\Omega_0 t}{2} \right) \cos \left(\frac{\Omega_0 t}{2} \right) = \frac{i}{2} \sin (\Omega_0 t) \tag{4.11}$$

These Rabi frequencies (Eqs. 4.8, 4.10) are analogous to the coupling of two spin states by an oscillating *magnetic* field (see Appendix 4.D). Equations 4.6 and 4.7 constitute the solutions to the optical Bloch equations for a two-level system. They describe the time-evolution of the populations and coherences of a two-level atom coupled by a single-mode optical field. However they do not include spontaneous emission, an omission that we address in Section 4.5.

Equations 4.5 were obtained in the *interaction representation* of the two-level atom density matrix in which the time evolution of the system is driven by the time dependence of the coupling operator, $\hat{V}(t) = \hbar\Omega_0 \cos\omega t$. With the help of Eq. 3.34, we can see that in order to switch to the *Schrödinger representation*, we can invoke the transformation, $\tilde{\rho}_{12} = \rho_{12} e^{i\omega_0 t}$. The result is

$$\frac{d\tilde{\rho}_{12}}{dt} = i\omega\tilde{\rho}_{12} + \frac{i}{\hbar}[V_{12} e^{i\omega t}(\tilde{\rho}_{11} - \tilde{\rho}_{22})]$$

and we see that the time dependence of the coherence matrix element contains the extra $i\omega_0\tilde{\rho}_{12}$ on the right, as in Eq. 3.44.

4.3 Atom Bloch vector

Following Eqs. 3.50 and 3.51 we can write the time-dependence of the atom density matrix elements in the form of a matrix equation,

$$\begin{bmatrix} \frac{d\rho_{11}}{dt} \\ \frac{d\rho_{12}}{dt} \\ \frac{d\rho_{21}}{dt} \\ \frac{d\rho_{22}}{dt} \end{bmatrix} = \frac{i}{\hbar} \begin{bmatrix} 0 & \frac{\hbar\Omega_0}{2} & -\frac{\hbar\Omega_0}{2} & 0 \\ \frac{\hbar\Omega_0}{2} & -\hbar\Delta\omega & 0 & -\frac{\hbar\Omega_0}{2} \\ -\frac{\hbar\Omega_0}{2} & 0 & \hbar\Delta\omega & \frac{\hbar\Omega_0}{2} \\ 0 & -\frac{\hbar\Omega_0}{2} & \frac{\hbar\Omega_0}{2} & 0 \end{bmatrix} \begin{bmatrix} \rho_{11} \\ \rho_{12} \\ \rho_{21} \\ \rho_{22} \end{bmatrix}$$

or as a vector product involving the Bloch vector $\boldsymbol{\beta}$ and the torque vector $\boldsymbol{\Omega}$, first introduced in Section 3.2.6

$$\frac{d\boldsymbol{\beta}}{dt} = -\boldsymbol{\beta} \times \boldsymbol{\Omega}$$

where the Bloch vector β can be expressed in terms of the circular or Cartesian matrix elements (see Appendix 4.A) of the atom density matrix as

$$\begin{aligned} \boldsymbol{\beta} &= \hat{\imath}\,(\rho_{21} + \rho_{12}) + \hat{\jmath}\,(\rho_{21} - \rho_{12}) + \hat{\mathbf{k}}\,(\rho_{22} - \rho_{11}) \\ &= \hat{\imath}\left(\langle\sigma^-\rangle + \langle\sigma^+\rangle\right) + \hat{\jmath}i\left[\langle\sigma^-\rangle - \langle\sigma^+\rangle\right] + \hat{\mathbf{k}}\left(\langle\sigma^+\sigma^-\rangle - \langle\sigma^-\sigma^+\rangle\right) \\ &= \hat{\imath}\,\langle\sigma_x\rangle + \hat{\jmath}\,\langle\sigma_y\rangle + \hat{\mathbf{k}}\,\langle\sigma_z\rangle \end{aligned} \tag{4.12}$$

with

$$\frac{d\boldsymbol{\beta}}{dt} = \hat{\imath}\Delta\omega\,\langle\sigma_y\rangle - \hat{\jmath}\left[\Omega_0\,\langle\sigma_z\rangle + \Delta\omega\,\langle\sigma_x\rangle\right] + \hat{\mathbf{k}}\Omega_0\,\langle\sigma_y\rangle$$

and the torque vector $\boldsymbol{\Omega}$ is written as

$$\boldsymbol{\Omega} = \frac{1}{\hbar}\left[\hat{\imath}\hbar\Omega_0 - \hat{\mathbf{k}}\hbar\Delta\omega\right] \tag{4.13}$$

Note that the length of the torque vector is just the Rabi frequency first introduced in Section 4.2:

$$\left|(\boldsymbol{\Omega})^2\right|^{1/2} = \Omega = \sqrt{\Omega_0^2 + (\Delta\omega)^2}$$

Now with the expressions for the three time-dependent components of the Bloch vector

$$\begin{aligned} \frac{d\beta_x}{dt} &= \Delta\omega\,\langle\sigma_y\rangle \\ \frac{d\beta_y}{dt} &= -\Omega_0\,\langle\sigma_z\rangle - \Delta\omega\,\langle\sigma_x\rangle \\ \frac{d\beta_z}{dt} &= \Omega_0\,\langle\sigma_y\rangle \end{aligned}$$

and with the initial condition that at time $t = 0$, $\beta_x(0) = 0$, $\beta_y(0) = 0$, $\beta_z = -\beta$, we can solve for the time evolution of the Bloch vector components:

$$
\begin{aligned}
\beta_x &= |\beta| \frac{\Omega_0 \Delta\omega}{\Omega^2} (\cos \Omega t - 1) \\
\beta_y &= |\beta| \frac{\Omega_0}{\Omega} \sin \Omega t \\
\beta_z &= -|\beta| \left[1 + \left(\frac{\Omega_0}{\Omega} \right)^2 (\cos \Omega t - 1) \right]
\end{aligned}
\tag{4.14}
$$

The time dependence of the three components of the atom Bloch vector provides a useful illustration of the atom–field interaction. On-resonance coupling, $\Delta\omega = 0, \Omega = \Omega_0$, is the easiest to describe, with the situation depicted in Fig. 4.1. From Eqs. 4.14 we see that the Bloch vector initially points in the $-z$ direction, which from Eq. 4.12 obviously means that all the population is in the ground state. As time advances, the Bloch vector begins to rotate counterclockwise in the z–y plane At $t = \pi/2\Omega_0$ the Bloch vector is aligned along $+y$, and at $t = \pi/\Omega_0$ it points upward along $+z$. All the population has been transferred to the excited state. The Bloch vector continues to rotate (or *nutate)* about the torque vector $\mathbf{\Omega}$ (which, as can be seen from Eq. 4.13, points along $+x$ when $\Delta\omega = 0$) with a frequency proportional to the strength of the atom–field coupling through Ω_0. From Eq. 4.9 we see that the population oscillates between the ground- and excited-states with a frequency $\Omega_0/2$ as the energy $\hbar\omega_0$ alternately exchanges between the atom and the field. A resonant pulse of light of duration such that $\tau = \pi/2\Omega_0$ is called a "pi-over-two-pulse." After a $\pi/2$ pulse the difference between the excited and ground state population is zero and the time-dependent state function has equal components of each stationary state:

$$
\psi(t) = \cos\left(\frac{\Omega_0}{2} t\right) \psi_1 + \sin\left(\frac{\Omega_0}{2} t\right) \psi_2 \longrightarrow \frac{1}{\sqrt{2}} [\psi_1 + \psi_2]
$$

The equal mixing of ground and excited states results in a wavefunction with maximal transition moment, and we remember from Eq. 2.22 that the rate of spontaneous emission increases with the square of transition dipole moment. Now, if we consider an ensemble of atoms sufficiently dilute such that we can neglect collisional (irreversible) decoherence but sufficiently dense such that the mean distance between atoms is less than a resonance wavelength, then the transition dipoles of the individual atoms will couple to produce an ensemble dipole moment. If a $\pi/2$ pulse is applied to this ensemble, whose members are all initially in the ground state, the collective Bloch vector will nutate to $+y$ as in the case of the single atom. However, inhomogeneous broadening due to the thermal motion of the atoms will lead to subsequent dispersion of the individual atom Bloch vectors in the x–y plane. The time evolution of the collective Bloch

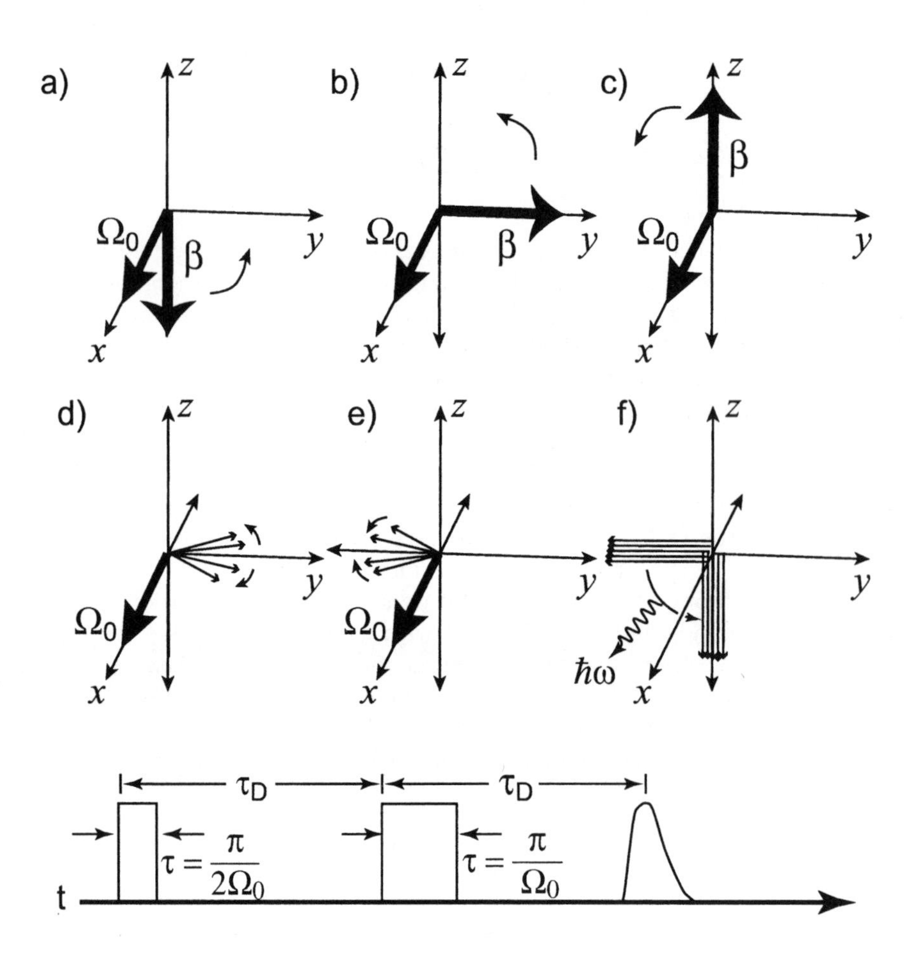

Figure 4.1: Panels (a),(b),(c) show precession of Bloch vector about torque vector ($\Delta\omega = 0$) for a single atom. Panels (d), (e), (f) show Bloch vector for an *ensemble* of atoms rotated to $+y$ axis with $\pi/2$ pulse, followed by inhomogeneous broadening interval τ_D, π pulse, superradiant photon echo relaxation to ensemble ground state.

vector after the $\pi/2$ pulse will be

$$\begin{aligned}\beta_x &= -|\boldsymbol{\beta}|\sin\Delta\omega t\\ \beta_y &= |\boldsymbol{\beta}|\cos\Delta\omega t\\ \beta_z &= 0\end{aligned}$$

where $\Delta\omega$ reflects the inhomogeneous phase dispersion. If now after a time τ a π pulse is applied to the ensemble, the distributed Bloch vectors will undergo a phase advance of $\pi - 2\Delta\omega\tau$ and continue to time-evolve as

$$\begin{aligned}\beta_x &= |\boldsymbol{\beta}|\sin\left[\Delta\omega\,(t-\tau)\right]\\ \beta_y &= -|\boldsymbol{\beta}|\cos\left[\Delta\omega\,(t-\tau)\right]\\ \beta_z &= 0\end{aligned}$$

After a second time interval τ the individual Bloch vectors will all point toward $-y$ and the collective transition dipole will again be maximal. The phasing of the individual dipoles produces a cooperative spontaneous emission from the ensemble, which is called the *photon echo.* The signature of the photon echo is twofold: (1) the appearance of a pulse of fluorescence after a delay τ from the end of the applied π pulse and (2) a fluorescence rate varying as the *square* of the excited state population. This unusual behavior arises from the individual dipole coupling and results in rapid depopulation of the excited state with a fluorescence lifetime much shorter than that of the individual atoms. This collective phasing of individual dipoles is called *superradiance.* It is important to bear in mind that the photon echo does not illustrate a recovery of coherence from an irreversible process. It works only for inhomogeneous broadening, due to a well-defined distribution of atomic kinetic energy, in which the time evolution of the individual members of the atom ensemble have not undergone random phase interruptions.

4.4 Preliminary Discussion of Spontaneous Emission

4.4.1 Susceptibility and polarization

Everything we have developed up to this point involves the coupling of one optical field mode to a two-level atom. In fact for this situation the Schrödinger equation is perfectly adequate to describe the time evolution of the system because it can always be described by a wavefunction, that is, a pure state. With the inclusion of spontaneous emission, the system can be described only by a probability distribution of final states, and therefore the density matrix description becomes indispensable.

Equations 2.10 and 2.11 do not take into account the fact that the excited state is coupled to all the supposedly empty modes of the radiation field as well as to the applied laser frequency ω. In order to take spontaneous emission into account we will first go back to Section 1.5 to recalculate the absorption coefficient K (Eqs. 1.43, 1.59) starting from the relation between the susceptibility and the polarization, Eq. 1.37. In order to get a new expression for the susceptibility, we will write the polarization in terms of a collection of individual two-level transition dipoles. We will use the solutions for the coefficients of our coupled two-level atom (Eqs. 2.10, 2.11). However, we will modify the expression for C_2 by adding a term that reflects the spontaneous emission of the upper state. The resulting expression for the susceptibility (and therefore the absorption coefficient) will reflect the finite "natural" lifetime of the upper state. For the present discussion we are concerned only with the time-dependence of the real optical wave which we express as,

$$\mathbf{E}(t) = \mathbf{E}_0 \cos\omega t = \frac{1}{2}\mathbf{E}_0 \left[e^{i\omega t} + e^{-i\omega t}\right]$$

and then consider how to write the polarization in terms of the susceptibility when the field contains the two conjugate frequencies, $\pm\omega$. Substituting in Eq. 1.37, we have

$$\mathbf{P}(t) = \frac{1}{2}\epsilon_0 \mathbf{E}_0 \left[\chi(\omega)e^{i\omega t} + \chi(-\omega)e^{-i\omega t}\right] \tag{4.15}$$

The polarization can also be expressed in terms of the density of transition dipoles in a gas of two-level atoms

$$\mathbf{P}(t) = \frac{N}{V}\mathbf{d}_{12}(t) \longrightarrow \frac{N}{V}\langle \mathbf{d}_{12} \rangle \tag{4.16}$$

where $\mathbf{d}$ is the transition dipole of a single atom, N/V is the atom density, and the quantum-mechanical expectation value for the transition dipole moment is the vector version of Eq. 2.7:

$$\langle \mathbf{d}_{12} \rangle = -e \int \Psi \sum_j \mathbf{r_j} \Psi d\tau$$

We will use the interpretation of the polarization $\mathbf{P}$ as the density of transition dipoles extensively in the theory of the laser (see Chapter 7, Section 7.2.1). Now, from Eq. 2.3

$$\langle \mathbf{d}_{12} \rangle = -e \left[C_1^* C_2 \left\langle \psi_1 \left| \sum_j \mathbf{r_j} \right| \psi_2 \right\rangle e^{-i\omega_0 t} + C_2^* C_1 \left\langle \psi_2 \left| \sum_j \mathbf{r_j} \right| \psi_1 \right\rangle e^{i\omega_0 t} \right]$$

To make the notation less cumbersome, we define

$$\langle \mathbf{r}_{mn} \rangle = \left\langle \psi_m \left| \sum_j \mathbf{r_j} \right| \psi_n \right\rangle$$

so then we have

$$\langle \mathbf{d}_{12} \rangle = -e \left[C_1^* C_2 \langle \mathbf{r}_{12} \rangle e^{-i\omega_0 t} + C_2^* C_1 \langle \mathbf{r}_{21} \rangle e^{i\omega_0 t} \right] \tag{4.17}$$

In principle all we have to do is substitute the solutions for the coupled equations relating C_1, C_2 from Eqs. 2.10 and 2.11 into Eq. 4.17, which in turn can be inserted into Eq. 4.16 to obtain an expression for the polarization in terms of atomic properties and the driving field. However, the solution for C_2, Eq. 2.13, does not take into account spontaneous emission. We are now going to make an ad hoc modification of Eq. 2.11 to include a radiative loss rate constant γ:

$$\Omega_0^* \cos \omega t \, e^{i\omega_0 t} C_1 - i\gamma C_2 = i\frac{dC_2}{dt} \tag{4.18}$$

This term by no means "explains" spontaneous emission. It simply acknowledges the existence of the effect and characterizes its magnitude by γ. If the coupling field is shut off ($\Omega_0^* = 0$)

$$-i\gamma C_2 = i\frac{dC_2}{dt}$$

and

$$C_2(t) = C_2(t=0)e^{-\gamma t}$$

Now the probability of finding the atom in the excited state is

$$|C_2(t)|^2 = |C_2(t=0)|^2 \, e^{-2\gamma t}$$

and the number of atoms N_2 in the excited state of an ensemble N is

$$N_2 = N\,|C_2(t)|^2 = N_2^0 e^{-2\gamma t}$$

where N_2^0 is the number of excited-state atoms at $t = 0$. If we compare this behavior to the result obtained from the Einstein rate expression, Eq. 1.30, we see immediately that[1]

$$A_{21} = 2\gamma \equiv \Gamma \tag{4.19}$$

Now the steady-state solution for our new, improved $C_2(t)$ coefficient is

$$C_2(t) = -\frac{1}{2}\Omega_o^* \left[\frac{e^{i(\omega_0+\omega)t}}{\omega_0 + \omega - i\gamma} + \frac{e^{i(\omega_0-\omega)t}}{\omega_0 - \omega - i\gamma}\right]$$

and we take the weak-field approximation for $C_1(t) \simeq 1$. These values for C_1, C_2 substituted back into Eq. 4.17 for transition dipole yield

$$\langle \mathbf{d}_{12} \rangle = \frac{e^2\,|\langle \mathbf{r}_{12}\rangle|^2\,\mathbf{E}_0}{2\hbar}\left[\frac{e^{i\omega t}}{\omega_0 + \omega - i\gamma} + \frac{e^{-i\omega t}}{\omega_0 - \omega - i\gamma} + \frac{e^{-i\omega t}}{\omega_0 + \omega + i\gamma} + \frac{e^{i\omega t}}{\omega_0 - \omega + i\gamma}\right] \tag{4.20}$$

[1]Note that it is customary in laser theory (Chapter 7) to use Γ for the sum over *all* dissipative processes (spontaneous emission, collisions, etc.)

which in turn we insert in Eq. 4.16. After replacing $|\langle \mathbf{r}_{12} \rangle|^2$ with its orientation-averaged value, $\frac{1}{3}|\langle \mathbf{r}_{12} \rangle|^2$, we have for the polarization vector

$$\mathbf{P}(t) = \frac{N}{V}\frac{e^2}{6\hbar}|\langle \mathbf{r}_{12} \rangle|^2 \mathbf{E}_0 \left[\left(\frac{1}{\omega_0 - \omega - i\gamma} + \frac{1}{\omega_0 + \omega + i\gamma}\right) e^{-i\omega t} + \right. \tag{4.21}$$

$$\left. \left(\frac{1}{\omega_0 + \omega - i\gamma} + \frac{1}{\omega_0 - \omega + i\gamma}\right) e^{i\omega t}\right] \tag{4.22}$$

Comparing this result to Eq. 1.37, we identify $\chi(\omega)$, the susceptibility in terms of the atomic properties and the driving field frequency:

$$\chi(\omega) = \frac{Ne^2 |\langle \mathbf{r}_{12} \rangle|^2}{3\epsilon_0 \hbar V}\left(\frac{1}{\omega_0 - \omega - i\gamma} + \frac{1}{\omega_0 + \omega + i\gamma}\right) \tag{4.23}$$

Separating the real and imaginary parts, we have

$$\chi(\omega) = \frac{Ne^2 |\langle \mathbf{r}_{12} \rangle|^2}{3\epsilon_0 \hbar V}\left[\left(\frac{\omega_0 - \omega}{(\omega_0 - \omega)^2 + \gamma^2} + \frac{\omega_0 + \omega}{(\omega_0 + \omega)^2 + \gamma^2}\right)\right.$$
$$\left. + i\gamma\left(\frac{1}{(\omega_0 - \omega)^2 + \gamma^2} - \frac{1}{(\omega_0 + \omega)^2 + \gamma^2}\right)\right] \tag{4.24}$$

In any practical laboratory situation ω will never be more than several hundred gigaherz detuned from ω_0 so $|\omega_0 - \omega| \lesssim 10^{11}$ Hz. Since optical frequencies $\omega \sim 10^{15}$ Hz, it is clear that the second term on the right side of Eq. 4.23 will be negligible compared to the first term. Therefore we can drop the second term and write the susceptibility as

$$\chi(\omega) \cong \frac{Ne^2 |\langle \mathbf{r}_{12} \rangle|^2}{3\epsilon_0 \hbar V}\left(\frac{1}{\omega_0 - \omega - i\gamma}\right) \tag{4.25}$$

$$= \frac{Ne^2 |\langle \mathbf{r}_{12} \rangle|^2}{3\epsilon_0 \hbar V}\frac{\omega_0 - \omega + i\gamma}{(\omega_0 - \omega)^2 + \gamma^2} \tag{4.26}$$

$$= \frac{N\mu_{12}^2}{3\epsilon_0 \hbar V}\left(-\frac{\Delta\omega}{(\Delta\omega)^2 + (\Gamma/2)^2} + i\frac{\Gamma/2}{(\Delta\omega)^2 + (\Gamma/2)^2}\right) \tag{4.27}$$

$$= \frac{n\hbar\Omega_0^2}{3\varepsilon_0 E_0^2}\left(-\frac{\Delta\omega}{(\Delta\omega)^2 + (\Gamma/2)^2} + i\frac{\Gamma/2}{(\Delta\omega)^2 + (\Gamma/2)^2}\right) \tag{4.28}$$

Identifying the real and imaginary parts

$$\chi(\omega) = \chi' + i\chi''$$

we can, from Eq. 1.43, finally express the absorption coefficient as

$$K = \frac{\omega}{c\eta}\chi''(\omega) = \frac{\pi N\mu_{12}^2\omega_0}{3\epsilon_0 \hbar c V}\frac{\Gamma/2}{\pi\left[(\Delta\omega)^2 + (\Gamma/2)^2\right]} \tag{4.29}$$

The factor

$$\frac{\Gamma/2}{\pi\left[(\Delta\omega)^2+(\Gamma/2)^2\right]}=L(\omega-\omega_0)$$

is our familiar Lorentzian line-shape factor, and it governs the frequency dependence of the absorption coefficient. We see that K exhibits a peak at the resonance frequency ω_0 and a width of Γ. The factor of π inserted into the numerator and denominator of the right member of Eq. 4.29 permits normalization of the line-shape factor

$$\int_{-\infty}^{\infty}\frac{\Gamma/2}{\pi\left[(\Delta\omega)^2+(\Gamma/2)^2\right]}d\omega=1$$

We have also assumed in Eq. 4.29 that the gas is sufficiently dilute that $\eta\simeq 1$ and that the line shape is sufficiently narrow to replace ω with ω_0 so that

$$\frac{\omega}{c\eta}\longrightarrow\frac{\omega_0}{c}$$

The absorption cross section also exhibits the same line shape since from Eqs. 1.59 and 4.29 we have

$$\sigma_{0a}=\frac{\pi\mu_{12}^2\omega_0}{3\epsilon_0\hbar c}\frac{\Gamma/2}{\pi\left[(\Delta\omega)^2+(\Gamma/2)^2\right]}\tag{4.30}$$

consistent with our earlier expression for the frequency dependence of the absorption cross section, Eq. 2.34. We can also write the imaginary component of the susceptibility in terms of the cross section using Eqs. 1.59 and 4.29

$$\chi''=\frac{c}{\omega_0}\frac{N}{V}\sigma_{0a}\tag{4.31}$$

4.4.2 Susceptibility and the driving field

At moderate intensities much of the physics of atom–light–field interaction can be gleaned from the simple model of a harmonically bound electron driven by an external classical oscillating field. We illustrate the use of this "driven charged oscillator" model in this section. We shall see it again when we discuss optical cooling and trapping (Chapter 6), and it is the underlying model of most of laser theory (Chapter 7). Let us return to the expression of the polarization in terms of the susceptibility (Eq. 4.15)

$$\mathbf{P}(t)=\frac{1}{2}\epsilon_0\mathbf{E}_0\left[\chi(\omega)e^{i\omega t}+\chi(-\omega)e^{-i\omega t}\right]$$

with

$$\chi(\omega)=\chi'+i\chi''$$

Substitution of the real and imaginary parts of the susceptibility into the polarization produces

$$\mathbf{P}(t) = \frac{1}{2}\epsilon_0 \mathbf{E}_0 \left\{ \left[\chi'(\omega) + i\chi''(\omega)\right] e^{i\omega t} + \left[\chi'(-\omega) + i\chi''(-\omega)\right] e^{-i\omega t} \right\} \tag{4.32}$$

Equation 4.24 shows that the real part of the susceptibility is symmetric in ω while the imaginary part is antisymmetric

$$\chi'(-\omega) = \chi'(\omega)$$
$$\chi''(-\omega) = -\chi''(\omega)$$

so that the real polarization can be written

$$\mathbf{P_r}(t) = \epsilon_0 \mathbf{E}_0 \left[\chi'(\omega) \cos \omega t - \chi''(\omega) \sin \omega t\right] \tag{4.33}$$

Equation 4.33 shows that the real, dispersive part of the susceptibility is in phase with the driving field while the imaginary, absorptive part follows the driving field in quadrature. As the optical field drives the polarizable atom, we can examine the steady-state energy flow between the driving field and the driven atom. The polarization $\mathbf{P}(t)$ is just the density of an ensemble of dipoles:

$$\mathbf{P}(t) = \left(\frac{N}{V}\right) \mathbf{d}_{12} \tag{4.34}$$

This polarization interacts back with the light field that produced it. Imagine that we have a linearly polarized light beam of circular cross section with radius $\mathbf{r}$, frequency ω, with well-defined (Gaussian) "edges" in the transverse plane, but propagating along z as a plane wave. Later (see Chapter 8 Section 8.4) this beam will be called the *fundamental Hermite-Gaussian beam.* We write the traveling wave in its complex form as

$$\mathbf{E}(\mathbf{r}, z, t) = \mathbf{E}_0(\mathbf{r}) e^{i(kz-\omega t)} \tag{4.35}$$

and the complex polarization[2] as

$$\mathbf{P} = \epsilon_0 \chi \mathbf{E} = \epsilon_0(\chi' + i\chi'')\mathbf{E}_0 e^{i(kz-\omega t)} \tag{4.36}$$

We now write the polarization as the sum of a dispersive component and an absorptive component

$$\mathbf{P} = \mathbf{P}_{\text{dis}} + \mathbf{P}_{\text{abs}} \tag{4.37}$$

with

$$\mathbf{P}_{\text{dis}} = \epsilon_0 \chi' E_0 e^{i(kz-\omega t}$$

and

$$\mathbf{P}_{\text{abs}} = i\epsilon_0 \chi'' E_0 e^{i(kz-\omega t)}$$

[2]Note that the real and complex forms of the polarization are related by $\mathbf{P_r} = \frac{1}{2}\left(\mathbf{P} + \mathbf{P}^*\right)$. We will see these forms again in the theory of the laser, Chapter 7, Section 7.2.1.

The energy density within a transparent dielectric, isotropic material with no permanent dipole moment, interacting with the electric field of this light beam is given by

$$\begin{aligned} \mathcal{E}_{\text{dis}}(t) &= -\text{Re}[\mathbf{P}] \cdot \text{Re}[\mathbf{E}^*] \\ &= -\epsilon_0 E_0^2 \left[\chi' \cos^2(kz - \omega t) - \chi'' \sin(kz - \omega t) \cos(kz - \omega t)\right] \end{aligned}$$

Optical cycle averaging yields

$$\langle \mathcal{E} \rangle_{\text{dis}} = -\frac{1}{2}\,\epsilon_0 E_0^2(\mathbf{r}) \chi'(\omega) \tag{4.38}$$

Equation 4.38 should be interpreted as the energy associated with a collection of driven atom transition dipoles interacting with the driving **E**-field. Since the polarization is a density of dipoles, the interaction energy is really an energy density.

We can write an expression for the light *force* acting on this collection of transition dipoles by first taking the spatial gradient of the interaction energy and then again taking the optical cycle average:

$$\begin{aligned} \mathbf{F}_{\text{dis}} &= \text{Re}\left[-\boldsymbol{\nabla}\left(-\mathbf{P}_{\text{dis}} \cdot \mathbf{E}^*\right)\right] \\ &= \text{Re}\left[\mathbf{P}_{\text{dis}} \boldsymbol{\nabla} \mathbf{E}^*\right] && (4.39) \\ &= \frac{1}{2}\,\epsilon_0 \chi' \boldsymbol{\nabla} E_0^2 && (4.40) \\ & && (4.41) \end{aligned}$$

Averaging over the optical cycle, we obtain

$$\begin{aligned} \langle \mathbf{F}_{\text{dis}} \rangle &= \frac{1}{2}\,\text{Re}\left[\mathbf{P}_{\text{dis}}\right] \text{Re}\left[\boldsymbol{\nabla} \mathbf{E}^*\right] && (4.42) \\ &= \frac{1}{4}\,\epsilon_0 \chi' \boldsymbol{\nabla} E_0^2(\mathbf{r}) && (4.43) \end{aligned}$$

The spatial gradient of the **E**-field is in the transverse plane of the propagating light wave. The direction of the force depends on the sign of χ', the dispersive part of the polarization, and the sign of the field gradient. If the light beam is tuned to the red of resonance, Eq. 4.28 shows that χ' is positive, and the force is in the same direction as the gradient that is negative in the transverse plane. The atoms will be attracted transversely toward the interior of the light beam where the field is highest. Along the longitudinal direction the field gradient (and therefore the force) is negligible so the atoms are free to drift along z while being constrained transversely. We will see in Chapter 6 that a potential can be derived from this "dipole-gradient" force, so we can think of the light beam as providing an attractive potential tube along which the atoms can be transported. Tuning to the blue reverses the force sign, and the atoms will be ejected from the light beam. Field gradients can also be created by focusing a laser beam, by standing light waves, or by generating evanescent fields near dielectric surfaces.

If the light is tuned very near a resonance, the energy of the driving field will be absorbed. We therefore write this absorptive interaction energy as

$$\mathcal{E}_{\text{abs}} = -\text{Im}[\mathbf{P}] \cdot \text{Re}[\mathbf{E}^*] \tag{4.44}$$

and the cycle average

$$\langle \mathcal{E} \rangle_{\text{abs}} = -\frac{1}{2}\,\epsilon_0 \chi''(\omega) E_0^2 \tag{4.45}$$

We can associate a light force with this absorbed energy as well,

$$\begin{aligned} \mathbf{F}_{\text{abs}} &= \text{Re}\left[-\boldsymbol{\nabla}\left(-\mathbf{P}_{\text{abs}} \cdot \mathbf{E}^*\right)\right] \\ &= \text{Re}\left[\mathbf{P}_{\text{abs}} \boldsymbol{\nabla} \mathbf{E}^*\right] \end{aligned} \tag{4.46}$$

Taking the optical cycle average, we have

$$\langle \mathbf{F}_{\text{abs}} \rangle = \frac{1}{2}\,\text{Re}[\mathbf{P}][\text{Re}\boldsymbol{\nabla}\mathbf{E}^*] \tag{4.47}$$

$$= \frac{1}{2}\,\epsilon_0 \chi''(\omega) E_0^2 k \hat{\mathbf{k}} \tag{4.48}$$

where $\hat{\mathbf{k}}$ is the unit vector in the propagation direction. Here we consider the light beam as a plane wave of infinite transverse extent propagating along the z axis. The only spatial gradient therefore is in the phase of the traveling wave, and the force is in the direction of light - beam propagation. In taking the gradient of the interaction energy (Eqs. 4.39, 4.46), we have dropped the $E^* \,\boldsymbol{\nabla} P$ term. The reason is that from Maxwell's equations the polarization gradient of a neutral dielectric over spatial dimensions greater than atomic dimensions must be zero. Since the gradient of the field amplitudes extend over the dimensions of the light beam or, at the very smallest, the wavelength of light, the polarization gradient, whose characteristic scale length is of the order of the atomic dipole moment, can be safely ignored. Returning to Eq. 4.48, we see that the magnitude of this force depends on the light intensity along z (see Eq. 1.10) and the magnitude of χ'', proportional to the cross section for light absorption (see Eq. 4.31). This force is sometimes called the "radiation pressure" force. We will discuss it again in terms of the cross section for classical radiation of an oscillating electron in Appendix 7.A, Section 7.A.2. The atom absorbs light energy from the field and will reemit it by spontaneous emission. In fact, due to spontaneous emission, the magnitude of this force does not increase indefinitely with light intensity, but "saturates" when the rate of stimulated absorption becomes equal to the rate of spontaneous emission. Of course, both the dipole gradient and radiation pressure forces are present whenever a light beam of frequency ω passes through matter with susceptibility $\chi(\omega)$. Because of the frequency dependence of the dispersive and absorptive components of the susceptibility, however, (*v.s.* Eqs. 4.27, 4.28) the dipole gradient force dominates with light tuned far off-resonance and the radiation pressure force is most important with the light tuned within the natural width of the absorbing transition. Both the dipole gradient force and the radiation pressure force are of great importance for the cooling and manipulation of atoms. We will examine their properties in more detail in Chapter 6.

It is also worthwhile to consider how the average *power* of the field–atom interaction is distributed between the dispersive and absorptive parts of the susceptibility. The power density applied to the polarizable atom from the driving electric field is given by

$$\wp = \frac{d\mathbf{P}}{dt} \cdot \mathbf{E}(t)$$

We need only consider the time dependence of the light field, so we take $E(t) = E_0 \cos \omega t$, and the expression for $\wp$ becomes

$$\wp = \epsilon_0 E_0^2 \omega \left[\chi' \sin \omega t \cos \omega t - \chi'' \cos^2 \omega t \right]$$

Averaging over an optical cycle results in

$$\langle \wp \rangle = -\frac{1}{2} \epsilon_0 E_0^2 \omega \chi'' (\omega)$$

and again from Eq. 4.28 close to resonance

$$\langle \wp \rangle = -\frac{1}{3} \frac{\Omega_0^2}{\Gamma} n \hbar \omega_0$$

We see that the energy of the field flows to the absorptive part of the atomic response to the forced oscillation. Under conditions of steady-state excitation, the energy density flowing to an ensemble of atoms from the field must be balanced by the energy reradiated from the atoms. An ensemble of N classical dipoles in volume V, oscillating along a fixed direction, radiates energy density at the rate

$$\langle \Re \rangle = \frac{4\omega_0^4 \mu_{12}^2}{4\pi\epsilon_0 3c^3} \frac{N}{V} \tag{4.49}$$

and at steady state

$$\frac{1}{2} \epsilon_0 E_0^2 \omega_0 \chi'' = \frac{4\omega_0^4 \mu_{12}^2}{4\pi\epsilon_0 3c^3} \frac{N}{V}$$

or the incoming resonant energy flux absorbed must equal the flux radiated:

$$\frac{1}{2} \epsilon_0 E_0^2 c \chi'' = \frac{4\omega_0^3 \mu_{12}^2}{4\pi\epsilon_0 3c^2} \frac{N}{V}$$

Finally, from Eq. 4.31 we have

$$\sigma_{0a} = \frac{\frac{\omega_0^4 \mu_{12}^2}{4\pi\epsilon_0 3c^3}}{\frac{1}{2}\epsilon_0 E_0^2 c} = \frac{32\pi^3 \mu_{12}^2}{3\lambda^4 \epsilon_0^2 E_0^2} = \frac{\hbar\omega_0 A_{21}}{\frac{1}{2}\epsilon_0 c E_0^2}$$

which shows once again (see Eq.2.30) that the absorption cross section is simply the ratio of the power emitted to the incoming flux. Cross sections and rate equations figure importantly in the theory of the laser, and we shall have occasion to revisit the use of a "cross section" as an interaction strength parameter in Chapter 7.

4.5 Optical Bloch Equations with Spontaneous Emission

In order to find the optical Bloch equations including spontaneous emission, we insert the phenomenological $-i\gamma C_2$ term into Eq. 2.11 so that now we have

$$\Omega_0^* \cos(\omega t)\ e^{i\omega_0 t} C_1 - i\gamma C_2 = i\frac{dC_2}{dt}$$

and the resulting density matrix elements become

$$\frac{d\rho_{22}}{dt} = -i\frac{\Omega_0^*}{2}e^{-i(\Delta\omega)t}\rho_{12} + i\frac{\Omega_0}{2}e^{i(\Delta\omega)t}\rho_{21} - 2\gamma\rho_{22} \tag{4.50}$$

$$\frac{d\rho_{11}}{dt} = i\frac{\Omega_0^*}{2}e^{-i(\Delta\omega)t}\rho_{12} - i\frac{\Omega_0}{2}e^{i(\Delta\omega)t}\rho_{21} + 2\gamma\rho_{22} \tag{4.51}$$

$$\frac{d\rho_{12}}{dt} = i\frac{\Omega_0}{2}e^{i(\Delta\omega)t}(\rho_{11} - \rho_{22}) - \gamma\rho_{12} \tag{4.52}$$

$$\frac{d\rho_{21}}{dt} = -i\frac{\Omega_0^*}{2}e^{-i(\Delta\omega)t}(\rho_{11} - \rho_{22}) - \gamma\rho_{21} \tag{4.53}$$

The oscillatory factors are eliminated from Eqs. 4.50–4.53 by substituting $\widetilde{\rho}_{12}e^{i(\Delta\omega)t} = \rho_{12}$ and $\widetilde{\rho}_{21}e^{-i(\Delta\omega)t} = \rho_{21}$ with the resulting equations

$$\frac{d\rho_{22}}{dt} = -i\frac{\Omega_0^*}{2}\widetilde{\rho}_{12} + i\frac{\Omega_0}{2}\widetilde{\rho}_{21} - 2\gamma\rho_{22} = -\frac{d\rho_{11}}{dt} \tag{4.54}$$

$$\frac{d\widetilde{\rho}_{12}}{dt} = i\frac{\Omega_0}{2}(\rho_{11} - \rho_{22}) - \gamma\widetilde{\rho}_{12} - i(\Delta\omega)\widetilde{\rho}_{12} = \frac{d\widetilde{\rho}_{21}^*}{dt} \tag{4.55}$$

Now, setting the time derivatives to zero to get the steady-state solutions yields

$$\rho_{22} = \frac{\frac{1}{4}|\Omega_0|^2}{(\Delta\omega)^2 + \gamma^2 + \frac{1}{2}|\Omega_0|^2} \tag{4.56}$$

and

$$\rho_{12} = e^{i(\Delta\omega)t}\frac{\frac{1}{2}\Omega_0(\Delta\omega - i\gamma)}{(\Delta\omega)^2 + \gamma^2 + \frac{1}{2}|\Omega_0|^2} \tag{4.57}$$

We see that both the populations and coherences now have a frequency dependence with a Lorentzian denominator similar to but not identical with the Lorentzian line shapes we had previously found for the susceptibility χ, the absorption coefficient K, and the absorption cross section σ_{0a} (Eqs. 4.25, 4.29, 2.34). Now the denominators exhibit an extra $\frac{1}{2}|\Omega|^2$ term which makes the "effective" widths of ρ_{22} and ρ_{12}:

$$\Gamma_{\text{eff}} = 2\left(\left(\frac{\Gamma}{2}\right)^2 + \frac{1}{2}|\Omega_0|^2\right)^{1/2} \tag{4.58}$$

We can insert these new forms for ρ_{12} and $\rho_{21} = \rho_{12}^*$ from Eq. 4.57 into our previous expressions for the transition dipole $\langle\mu_{12}\rangle$ (Eqs. 4.17, 4.20) and then obtain new expressions for the polarization $\mathbf{P}(t)$ (Eqs. 4.16, 4.22); and the susceptibility χ (Eq. 4.27). The modified expression for the susceptibility is

$$\chi = \frac{N\mu_{12}^2}{3\epsilon_0 \hbar V}\left(-\frac{\Delta\omega}{(\Delta\omega)^2 + (\Gamma/2)^2 + \frac{1}{2}|\Omega_0|^2} + i\frac{\Gamma/2}{(\Delta\omega)^2 + (\Gamma/2)^2 + \frac{1}{2}|\Omega_0|^2}\right) \tag{4.59}$$

From the imaginary component of the susceptibility we obtain the new absorption coefficient

$$K = \frac{\omega}{c\eta}\chi''(\omega) = \frac{\pi N e^2 |\langle \mathbf{r}_{12}\rangle|^2 \omega_0}{3\epsilon_0 \hbar c V} \frac{\frac{\Gamma}{2\pi}}{(\Delta\omega)^2 + (\Gamma/2)^2 + \frac{1}{2}|\Omega_0|^2} \tag{4.60}$$

and the absorption cross section

$$\sigma_{0a} = \frac{\pi e^2 |\langle \mathbf{r}_{12}\rangle|^2 \omega_0}{3\epsilon_0 \hbar c} \frac{\frac{\Gamma}{2\pi}}{(\Delta\omega)^2 + (\Gamma/2)^2 + \frac{1}{2}|\Omega_0|^2} \tag{4.61}$$

The important new feature is the "effective width" Γ_{eff} which appears in χ, K, and σ_{0a}. Since $\Omega_0 = \boldsymbol{\mu}_{12}\cdot\mathbf{E}_0/\hbar$, it is clear that the effective width depends on the electric field amplitude and hence the intensity of the applied light field. The additional width of the absorption or emission line profile due to the intensity of the exciting light is called *power broadening.*

4.6 Mechanisms of Line Broadening

4.6.1 Power broadening and saturation

Equation 4.58 shows that as the power of the exciting light increases, the fractional population in the excited state saturates at a limiting value of $\rho_{22} = \frac{1}{2}$. This property is analogous to Eq. 1.45, which shows similar saturation behavior when the two-level atom is subject to broadband radiation. Note that Eqs. 4.59, 4.60, and 4.61, all with the same line shape factor, exhibit the same saturation characteristic. The *saturation parameter* defined by

$$S = \frac{\frac{1}{2}|\Omega_0|^2}{(\Delta\omega)^2 + (\Gamma/2)^2} \tag{4.62}$$

indexes the "degree of saturation". When the narrowband excitation light source is tuned to resonance, the saturation parameter is essentially a measure of the ratio of the on-resonance stimulated population transfer frequency Ω_0 to the spontaneous rate A_{21}. At resonance and with the saturation parameter equal to unity, we obtain

$$\Omega_0 = \frac{1}{\sqrt{2}}\Gamma \tag{4.63}$$

We can use Eq. 4.63 to define a "saturation power" I_{sat} for an atom with transition dipole μ_{12}. From Eq. 1.42 we have

$$E_0 = \sqrt{\frac{2\bar{I}}{\epsilon_0 c}}$$

so, using the conversion factor between μ_{12} and A_{21} entered in Table 2.1, we have

$$I_{\mathrm{sat}} = \frac{g_1}{g_2}\frac{2\pi^2 c\hbar}{3\tau\lambda_0^3} \tag{4.64}$$

A useful formula for practical calculations is

$$I_{\mathrm{sat}}(\mathrm{mW/cm^2}) = \frac{g_1}{g_2}\frac{2.081\times 10^{10}}{\tau(\mathrm{ns})\lambda_0^3(\mathrm{nm})}$$

Note that from Eqs. 4.56 and 4.63, using this definition of "saturation", $S=1$ and $\rho_{22}=\frac{1}{4}$. Some authors take the criterion for saturation to be $S=2$ in which case $\Omega_0=\Gamma$ and $\rho_{22}=\frac{1}{3}$.

Problem 4.1 *Calculate the saturation power I_{sat} for Na $3s\,^2S_{1/2} \longleftrightarrow 3p\,^2P_{3/2}$ and for Cs $6s\,^2S_{1/2} \longleftrightarrow 6p\,^2P_{3/2}$ in units of mW/cm^2*

4.6.2 Collision line broadening

The theory of atomic collisions covers a vast domain including elastic, inelastic, reactive, and ionizing processes. In low-pressure gases at ambient or higher temperature we need consider only the simplest processes: long-range van der Waals interactions that result in elastic collisions. The criterion of "low pressure" requires that the mean free path between collision be longer than any linear dimension of the gas volume. Collisions under these conditions can be modeled with straight-line trajectories during which the interaction time is short and the time between collisions is long compared to the radiative lifetime of the atomic excited state. Under these conditions the collisional interaction of the radiating atom can be characterized by a loss of coherence due to a phase interruption of the atomic excited-state wavefunction. The term "elastic" means that the collision does not affect the internal state populations so that we need consider only the off-diagonal elements of the density matrix

$$\frac{d\rho_{12}}{dt} = i\frac{\Omega_0}{2}e^{i(\omega-\omega_0)t}\left(\rho_{11}-\rho_{22}\right) - \gamma'\rho_{12}$$

where γ' is the sum[3] of the spontaneous emission γ and the collisional rate, γ_{col}

$$\gamma' = \gamma + \gamma_{\mathrm{col}} \tag{4.65}$$

[3]The reader is cautioned that the meaning of terms γ, γ', Γ can change with context. In an atomic physics context Γ usually means the spontaneous emission rate of an atom. In an engineering context Γ often denotes a phenomenological decay constant that is a sum over various decay processes (see Chapter 7). In the present context we are using γ' as sum of two identifiable decay processes.

and the inverse of the collision rate is just the time between phase interruptions or the time τ_{col} "between collisions." Now for hard-sphere collisions between atoms of mass m (with reduced mass $\mu = m/2$) and radius ρ in a single-species gas sample with density n, standard analysis of the kinetic theory of dilute gases shows that the time between collisions is

$$\tau_{\text{col}} = \frac{\sqrt{\frac{\mu}{\pi kT}}}{8\rho^2 n}$$

and the collision frequency is just

$$\tau_{\text{col}}{}^{-1} = \gamma_{\text{col}} = 8\rho^2 n \left(\frac{\pi kT}{\mu} \right)^{1/2} \tag{4.66}$$

Now we can relate this simple result of elementary gas kinetics to the rate of phase interruption by reinterpreting what we mean by the *collision radius.* When an excited atom, propagating through space, undergoes a collisional encounter, the long-range interaction will produce a time-dependent perturbation of the energy levels of the radiating atom and a phase shift in the radiation:

$$\eta = \int_{-\infty}^{\infty} [\omega(t) - \omega_0]\, dt = \int_{-\infty}^{\infty} \Delta\omega(t) dt$$

The long-range van der Waals interaction is expressed as

$$\Delta E = \hbar\Delta\omega = \frac{C_n}{\left[b^2 + (vt)^2\right]^{n/2}}$$

where b is the impact parameter of the collision trajectory and v is the collision velocity. The phase shift then becomes

$$\eta = \frac{1}{\hbar} \int_{-\infty}^{\infty} \frac{C_n}{\left[b^2 + (vt)^2\right]^{n/2}} dt$$

The integral is easily evaluated for the two most frequently encountered cases: $n = 6$ and $n = 3$, nonresonant and resonant van der Waals interactions, respectively. The phase shifts become

$$\eta_6(b) = \frac{2\pi}{3\hbar} \frac{C_6}{b_6^5 v}$$

and

$$\eta_3(b) = \frac{4\pi}{3\sqrt{3}\hbar} \frac{C_3}{b_3^2 v}$$

Now, if instead of using the hard-sphere criterion, we define a "collision" as an encounter that provokes at least a phase shift of unity, we have a new condition for the collision radius

$$b_6 = \left(\frac{2\pi}{3\hbar} \frac{C_6}{v} \right)^{1/5}$$

and

$$b_3 = \left(\frac{4\pi}{3\sqrt{3}\hbar}\frac{C_3}{v}\right)^{1/2}$$

and, taking the average collision velocity of a homogeneous gas sample at temperature T

$$v = \sqrt{\frac{8kT}{\pi\mu}}$$

we find the collision frequency γ_{col},

$$\gamma_{c6} = 4n\left(\frac{\sqrt{2}\pi^2 C_6}{3\hbar}\right)^{2/5}\left(\frac{4\pi kT}{\mu}\right)^{3/10}$$

$$\gamma_{c3} = 4n\left(\frac{2}{3}\right)^{3/2}\left(\frac{\pi^2 C_3}{\hbar}\right)$$

Substituting the generalized γ' from Eq. 4.65 for γ in the optical Bloch equations Eqs. 4.54 and 4.55 we find the steady-state solutions

$$\rho_{12} = e^{i(\omega-\omega_0)t}\frac{\frac{1}{2}\Omega_0(\omega-\omega_0-i\gamma')}{(\omega-\omega_0)^2+\gamma'^2+\frac{1}{2}\left(\frac{\gamma'}{\gamma}\right)|\Omega_0|^2}$$

and

$$\rho_{22} = \frac{\frac{1}{4}\left(\frac{\gamma'}{\gamma}\right)|\Omega_0|^2}{(\omega-\omega_0)^2+\gamma'^2+\frac{1}{2}\left(\frac{\gamma'}{\gamma}\right)|\Omega_0|^2}$$

The effective (radiative plus collision) line width becomes

$$\Gamma'_{\mathrm{eff}} = 2\left[\gamma'^2+\frac{1}{2}\left(\frac{\gamma'}{\gamma}\right)|\Omega_0|^2\right]^{1/2}$$

When the optical excitation is sufficiently weak that power broadening can be neglected compared to collision broadening, the second term on the right can be dropped, and the effective width becomes

$$\Gamma'_{\mathrm{eff}} = 2(\gamma+\gamma_{\mathrm{col}}) \tag{4.67}$$

Equations 4.58 and 4.67 express the limiting line widths for power broadening and collision broadening, respectively. Note that the susceptibility, absorption coefficient, and absorption cross section all retain the Lorentzian line shape, but with a width increased by the collision rate. Since every atom is subjected to the same broadening mechanism, collision broadening is an example of *homogeneous broadening.*

Problem 4.2 *At what pressure does the broadening due to collisions between ground-state sodium atoms equal the spontaneous emission line width of the resonance transition?*

4.6.3 Doppler broadening

Doppler broadening is simply the apparent frequency distribution of an ensemble of radiating atoms at temperature T. The radiation appears shifted because of the translational motion of the atoms. For each individual atom

$$\Delta\omega = \omega - \omega_0 = \mathbf{k}\cdot\mathbf{v} = kv_z$$

where $\mathbf{k}$ is the light-wave vector and $\mathbf{v}$ the atom velocity. This Doppler shift distribution of a gas ensemble in thermal equilibrium maps the Maxwell–Boltzmann probability distribution of velocities:

$$P(v_z)dv_z \sim e^{-\frac{mv_z^2}{2k_BT}}dv_z = e^{-\frac{mc^2(\Delta\omega)^2}{2\omega_0^2k_BT}}\frac{c}{\omega_0}d\omega \tag{4.68}$$

This distribution of frequencies is Gaussian with a peak at $\omega = \omega_0$ and a full width at half-maximum (FWHM) of

$$2\omega_0\left(\frac{2k_BT\ln 2}{mc^2}\right)^{1/2}$$

Another conventional measure of the width of this distribution is 2σ, twice the "standard deviation" used in the theory of the distribution $P(\varepsilon)$ of random measurement errors ε

$$P(\varepsilon) = \frac{1}{\sqrt{2\pi}\sigma}e^{-\varepsilon^2/2\sigma^2} \tag{4.69}$$

from which we can associate a spectral standard deviation:

$$2\sigma = \frac{2\omega_0}{c}\sqrt{\frac{k_BT}{m}}$$

The two measures of the width differ by a small factor:

$$\frac{\text{FWHM}}{2\sigma} = (2\ln 2)^{1/2} = 1.177$$

From Eqs. 4.68 and 4.69 we see that the normalized Doppler lineshape function is

$$D(\omega-\omega_0)d\omega = \frac{1}{\sqrt{2\pi}}\sqrt{\frac{m}{k_BT}}e^{-\frac{mc^2(\omega-\omega_0)^2}{2\omega_0^2k_BT}}d\omega \tag{4.70}$$

Figure 4.2 compares the Gaussian line shape of Eq. 4.70 to the Lorentzian line shape, Eq. 1.50

$$L(\omega-\omega_0)d\omega = \frac{\gamma}{2\pi}\frac{d\omega}{(\omega-\omega_0)^2 + \left(\frac{\gamma}{2}\right)^2} \tag{4.71}$$

associated with natural, power, and collision broadening. It is clear that for the two line shapes of equal width, the Gaussian profile dominates near line center and the Lorentzian is more important in the wings. Because the Doppler width is a property of the *ensemble* of atoms, with the Doppler shift of each atom having a unique but different value within the Maxwell-Boltzmann distribution , this type of broadening mechanism is called *heterogenous broadening.*

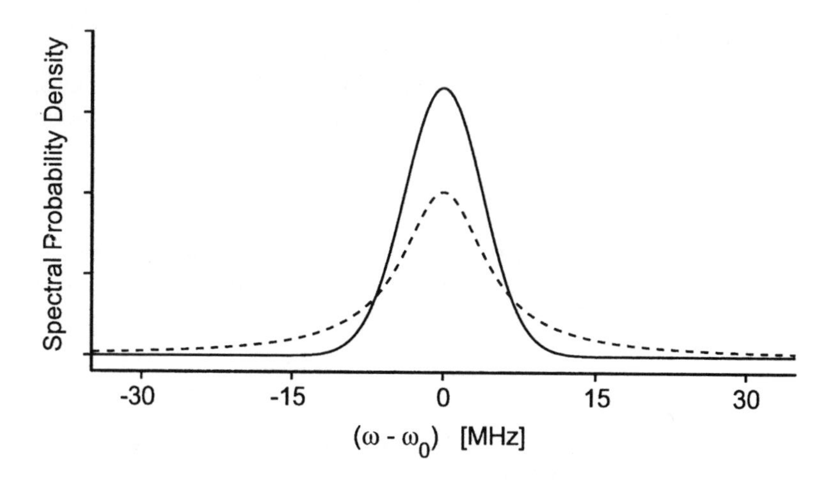

Figure 4.2: Spectral probability distribution (probability MHz^{-1}). The area under both curves normalized to unity. Gaussian distribution (solid line) and Lorentzian distribution (dashed line).

Problem 4.3 *Calculate the Doppler width for the resonance transition of an ensemble of sodium atoms at* 400° *C.*

4.6.4 Voigt profile

Of course in many practical circumstances both homogeneous and heterogeneous broadening contribute significantly to the line shape. In such cases we may consider that the radiation of each atom, homogeneously broadened by phase interruption processes such as spontaneous emission or collisions, is Doppler-shifted within the Maxwell-Boltzmann distribution at temperature T. The line profile of the gas ensemble must therefore be a convolution of the homogeneous and heterogenous line shapes. This composite line shape is called the *Voigt profile*:

$$\begin{aligned} V(\omega-\omega_0) &= \int_{-\infty}^{\infty} L(\omega-\omega_0-\omega')D(\omega-\omega_0)d\omega' \qquad (4.72) \\ &= \frac{\gamma}{2\sqrt{2\pi}\sigma}\int_{-\infty}^{\infty}\frac{e^{-(\omega-\omega_0)^2/2\sigma^2}}{(\omega-\omega_0-\omega')^2+\left(\frac{\gamma}{2}\right)^2}d\omega' \end{aligned}$$

Although there is no closed analytic form for this line shape, it is easily evaluated numerically.

Problem 4.4 *Calculate and plot the effective Lorentzian profile, the Gaussian profile, and the Voigt Profile for the resonance absorption line of sodium gas at a temperature of* 420° *C and a pressure of 700 mtorr.*

4.7 Further Reading

In this chapter we introduced spontaneous emission as a de facto loss term in the optical Bloch equations. A more serious treatment that gets most of the "right answer" is the Weisskopf–Wigner theory of spontaneous emission. The original reference is

- V. F. Weisskopf and E. Wigner, *Z. Phys.* **63**, 54 (1930).

but a more accessible discussion can be found in

- M. Sargent III, M. O.Scully, W. E. Lamb Jr., *Laser Physics*, Addison-Wesley, Reading, MA, 1974.

An updated and expanded discussion appears in

- M. O. Scully, M. S. Zubairy, *Quantum Optics*, Cambridge University Press, Cambridge, UK, 1997.

The time-dependence of the Bloch vector, superradiance, and photon echoes are treated in many engineering textbooks on quantum electronics and physics texts on quantum optics. Some examples of good treatments are

- A. Yariv, *Quantum Electronics*, 3rd edition, Wiley, New York, 1989.
- M. Sargent III, M. O. Scully, and W. E. Lamb, Jr.,*Laser Physics*, Addison-Wesley, Reading, MA, 1974.
- H. M. Nussenzveig,*Introduction to Quantum Optics*, Gordon & Breach, London, 1973.

Power broadening and elementary collision broadening are discussed in many places. In addition to the two books cited above, discussions can be found in

- R. Louden, *The Quantum Theory of Light*, 2nd edition, chapter 2, Clarendon Press, Oxford, 1983.
- M. Weissbluth, *Photon-Atom Interactions*, Chapter VI, Academic Press, Boston, 1989.
- A. Yariv, *Quantum Electronics*, 3rd edition, Wiley, New York, 1989.

For a deeper discussion of collision broadening and lineshape analysis,

- A. C. G Mitchell and M. W. Zemansky, *Resonance Radiation and Excited Atoms*, Cambridge University Press, Cambridge, UK, 1934.
- S. Y. Chen and M. Takeo, *Rev. Mod. Phys.* **29**, 20, (1957).
- R. E. M. Hedges, D. L. Drummond, and A. Gallagher, *Phys. Rev.* A **6**, 1519, (1972).
- A. Gallagher and T. Holstein, *Phys. Rev.* A **16** 2413, (1977).

Appendixes to Chapter 4

4.A Pauli Spin Matrices

In this appendix we illustrate the properties of the density operator applied to a spin $\frac{1}{2}$ system. We will see that the density matrix "toolbox" used to describe the two-level spin system is in fact applicable to any two-level problem and will help us analyze the unifying principles behind seemingly disparate physical phenomena.

We again start with a two-level system, but this time we imagine two states

$$|\alpha\rangle = \begin{pmatrix} 0 \\ 1 \end{pmatrix} \text{ and } |\beta\rangle = \begin{pmatrix} 1 \\ 0 \end{pmatrix}$$

that form a basis set spanning the space in which any arbitrary normalized state wavefunction may be expressed as

$$|\psi\rangle = a\,|\alpha\rangle + b\,|\beta\rangle\,; \qquad \langle\psi|\psi\rangle = |a|^2 + |b|^2 = 1$$

We have the usual orthonormal properties of the basis states

$$\langle\alpha|\alpha\rangle = \langle\beta|\beta\rangle = 1 \text{ and } \langle\alpha|\beta\rangle = 0$$

and now we introduce the *Pauli spin matrices* together with the identity matrix I

$$\sigma_x = \begin{bmatrix} 0 & 1 \\ 1 & 0 \end{bmatrix},\ \sigma_y = \begin{bmatrix} 0 & -i \\ i & 0 \end{bmatrix},\ \sigma_z = \begin{bmatrix} 1 & 0 \\ 0 & -1 \end{bmatrix},\ I = \begin{bmatrix} 1 & 0 \\ 0 & 1 \end{bmatrix} \tag{4.73}$$

and note that

$$\sigma_n^2 = I \qquad n = x, y, z$$

It is true that any 2×2 matrix can be represented by a linear combination of the Pauli spin matrices and I. For example, the 2×2 matrix representing the density operator ρ in the $|\alpha\rangle, |\beta\rangle$ space is,

$$\rho = \begin{bmatrix} \rho_{\alpha\alpha} & \rho_{\alpha\beta} \\ \rho_{\beta\alpha} & \rho_{\beta\beta} \end{bmatrix} = m_0 I + m_1\sigma_x + m_2\sigma_y + m_3\sigma_z \tag{4.74}$$

By inspection of the form of the Pauli spin matrices, Eq. 4.73, we can easily work out that

$$\rho = \begin{bmatrix} m_0 + m_3 & m_1 - im_2 \\ m_1 + im_2 & m_0 - m_3 \end{bmatrix} \tag{4.75}$$

and therefore from Eq. 3.11 we have

$$\mathrm{Tr}[\rho\sigma_x] = \langle\sigma_x\rangle = 2m_1 \tag{4.76}$$

$$\mathrm{Tr}[\rho\sigma_y] = \langle\sigma_y\rangle = 2m_2 \tag{4.77}$$

$$\mathrm{Tr}[\rho\sigma_z] = \langle\sigma_z\rangle = 2m_3 \tag{4.78}$$

$$\mathrm{Tr}\,[\rho I] = \langle I\rangle = 1 = 2m_0 \tag{4.79}$$

Substituting these values back into 4.75 gives us

$$\rho = \frac{1}{2}\begin{bmatrix} 1+\langle\sigma_z\rangle & \langle\sigma_x\rangle - i\langle\sigma_y\rangle \\ \langle\sigma_x\rangle + i\langle\sigma_y\rangle & 1-\langle\sigma_z\rangle \end{bmatrix} \tag{4.80}$$

and using Eq. 4.74 the density matrix can be expanded in terms of the Cartesian matrix elements and Pauli spin operators,

$$\rho = \frac{1}{2}\left[I + \langle\sigma_x\rangle\,\sigma_x + \langle\sigma_y\rangle\,\sigma_y + \langle\sigma_z\rangle\,\sigma_z\right] \tag{4.81}$$

Now it is evident from comparing Eqs. 4.74 and 4.80 that

$$\begin{aligned} \langle\sigma_z\rangle &= \rho_{\alpha\alpha} - \rho_{\beta\beta} \\ \langle\sigma_x\rangle &= \rho_{\alpha\beta} + \rho_{\beta\alpha} \\ \langle\sigma_y\rangle &= \frac{1}{i}\left(\rho_{\beta\alpha} - \rho_{\alpha\beta}\right) = i\left(\rho_{\alpha\beta} - \rho_{\beta\alpha}\right) \end{aligned} \tag{4.82}$$

The set of three matrices defined in Eq. 4.82 are often called the Cartesian components of the Pauli spin matrices. Notice that the average value of the z component of the Pauli spin operator represents the *population difference* between the excited- and ground-states of the two-level system.

In addition to the Cartesian spin matrices, often it is quite useful to introduce a new set of "circular" spin matrices $\sigma^+, \sigma^-, \sigma_z$ by defining σ^+, and σ^- as linear combinations of σ_x and σ_y

$$\begin{aligned} \sigma^+ &= \frac{1}{2}\left(\sigma_x + i\sigma_y\right) = \begin{bmatrix} 0 & 1 \\ 0 & 0 \end{bmatrix} \\ \sigma^- &= \frac{1}{2}\left(\sigma_x - i\sigma_y\right) = \begin{bmatrix} 0 & 0 \\ 1 & 0 \end{bmatrix} \end{aligned} \tag{4.83}$$

from which see we that

$$\begin{aligned} \sigma^+\sigma^- &= \begin{bmatrix} 1 & 0 \\ 0 & 0 \end{bmatrix} \\ \sigma^-\sigma^+ &= \begin{bmatrix} 0 & 0 \\ 0 & 1 \end{bmatrix} \end{aligned} \tag{4.84}$$

Note from Eqs. 4.83 that the x, y components of the spin matrix can be expressed in terms of the $+, -$ components as

$$\begin{aligned} \sigma_x &= \sigma^+ + \sigma^- \\ \sigma_y &= i\left(\sigma^- - \sigma^+\right) \end{aligned} \tag{4.85}$$

Inspection of the matrix for σ_z (Eq. 4.73) and Eqs. 4.84 allows us to write

$$\sigma_z = 2\sigma^+\sigma^- - I$$

Just as Eq. 4.80 expresses the density matrix in terms of the matrix elements $\langle\sigma_n\rangle$ with $n = x, y, z$, so we can express the density matrix in terms of matrix elements of the 'circular' components $\langle\sigma^+\rangle, \langle\sigma^-\rangle, \langle\sigma^+\sigma^-\rangle, \langle\sigma^-\sigma^+\rangle$

$$\rho = \begin{bmatrix} \langle\sigma^+\sigma^-\rangle & \langle\sigma^-\rangle \\ \langle\sigma^+\rangle & \langle\sigma^-\sigma^+\rangle \end{bmatrix} \tag{4.86}$$

with the expansion

$$\rho = \frac{1}{2}I + \left[\langle\sigma^+\sigma^-\rangle - \frac{1}{2}\right]\sigma_z + \langle\sigma^+\rangle\sigma^- + \langle\sigma^-\rangle\sigma^+ \tag{4.87}$$

It is worth noting the obvious but useful fact that

$$\langle\sigma^-\sigma^+\rangle = 1 - \langle\sigma^+\sigma^-\rangle$$

We will find that in various circumstances it will be convenient to express the density matrix either in terms of the Pauli Cartesian spin matrices (Eq. 4.81) or in terms of the "circular" spin matrices (Eq. 4.87).

Finally, notice that σ^+, σ^- have the interesting property of either provoking transitions between the levels of the spin $\frac{1}{2}$ system

$$\begin{aligned} \sigma^+ |\alpha\rangle &= \begin{bmatrix} 0 & 1 \\ 0 & 0 \end{bmatrix}\begin{pmatrix} 0 \\ 1 \end{pmatrix} = \begin{pmatrix} 1 \\ 0 \end{pmatrix} = |\beta\rangle \\ \sigma^- |\beta\rangle &= \begin{bmatrix} 0 & 0 \\ 1 & 0 \end{bmatrix}\begin{pmatrix} 1 \\ 0 \end{pmatrix} = \begin{pmatrix} 0 \\ 1 \end{pmatrix} = |\alpha\rangle \end{aligned} \tag{4.88}$$

or producing the null vector

$$\begin{aligned} \sigma^+ |\beta\rangle &= \begin{bmatrix} 0 & 1 \\ 0 & 0 \end{bmatrix}\begin{pmatrix} 1 \\ 0 \end{pmatrix} = \begin{pmatrix} 0 \\ 0 \end{pmatrix} \\ \sigma^- |\alpha\rangle &= \begin{bmatrix} 0 & 0 \\ 1 & 0 \end{bmatrix}\begin{pmatrix} 0 \\ 1 \end{pmatrix} = \begin{pmatrix} 0 \\ 0 \end{pmatrix} \end{aligned} \tag{4.89}$$

4.B Pauli Spin Matrices and Optical Coupling

We can express the Hamiltonian of our two-level atom in terms of the Pauli spin operators for the two-level system. We start with Eq. 2.2

$$\hat{H} = \hat{H}_A + \hat{V}$$

where we now write

$$\hat{H}_A = \frac{\hbar\omega_0}{2}\sigma_z$$

Then

$$\hat{H}_A |2\rangle = \frac{\hbar\omega_0}{2} |2\rangle$$

and

$$\hat{H}_A |1\rangle = -\frac{\hbar\omega_0}{2} |1\rangle$$

so that the energy difference between the two atomic levels is still

$$\Delta E = E_2 - E_1 = \hbar\omega_0$$

This choice for the energy levels allows us to write $\hat{H}_A$ as simply proportional to $\hat{\sigma}_z$, but it means that we must write the time-dependent state function for the two level atom as

$$\psi = \frac{1}{\sqrt{2}} \left[e^{i\frac{\omega_0}{2}t} |1\rangle + e^{-i\frac{\omega_0}{2}t} |2\rangle \right]$$

We write the light field as circularly polarized, propagating along the positive z axis with the electric field oscillating at frequency ω and rotating in the clockwise direction:

$$\hat{V} = \frac{\hbar\Omega_0}{2} \left[\sigma_x \cos\omega t + \sigma_y \sin\omega t\right]$$

Then from Eq. 4.85 we can write

$$\hat{V} = \frac{\hbar\Omega_0}{2} \left(e^{i\omega t}\sigma^- + e^{-i\omega t}\sigma^+ \right)$$

and taking matrix elements of $\hat{V}$ we find, using Eqs. 4.88 and 4.89,

$$V_{12} = \frac{\hbar\Omega_0}{2} e^{i\omega t} \quad (4.90)$$

$$V_{21} = \frac{\hbar\Omega_0}{2} e^{-i\omega t} \quad (4.91)$$

Now our two-level atom Hamiltonian has the following form, in terms of the Pauli spin matrices:

$$\hat{H} = \frac{\hbar}{2} \left[\omega_0\sigma_z + \Omega_0 \left(e^{i\omega t}\sigma^- + e^{-i\omega t}\sigma^+ \right) \right] \quad (4.92)$$

The matrix elements of this Hamiltonian operator in our two-level basis become

$$H = \frac{\hbar}{2} \begin{bmatrix} \omega_0 & \Omega_0 e^{i\omega t} \\ \Omega_0 e^{-i\omega t} & -\omega_0 \end{bmatrix}$$

4.C Time Evolution of the Optically Coupled Atom Density Matrix

Now that we have constructed the Hamiltonian in terms of the Pauli spin operators, Eq. 4.92, we insert it and the density matrix operator (Eq. 4.86) into the Liouville equation, Eq. 3.37, to obtain the time evolution of the density matrix.

The equations of motion for the time dependence of the atom density matrix can be written down directly from Eq. 3.41:

$$\begin{aligned}
\frac{d\rho_{11}}{dt} &= \frac{d}{dt}\left\langle\sigma^{+}\sigma^{-}\right\rangle = i\frac{\Omega_0}{2}\left[e^{-i\omega t}\left\langle\sigma^{-}\right\rangle - e^{i\omega t}\left\langle\sigma^{+}\right\rangle\right] \\
\frac{d\rho_{12}}{dt} &= \frac{d}{dt}\left\langle\sigma^{-}\right\rangle = i\omega_0\left\langle\sigma^{-}\right\rangle + i\frac{\Omega_0}{2}e^{i\omega t}\left[2\left\langle\sigma^{+}\sigma^{-}\right\rangle - 1\right] \\
\frac{d\rho_{21}}{dt} &= \frac{d}{dt}\left\langle\sigma^{+}\right\rangle = -i\omega_0\left\langle\sigma^{+}\right\rangle - i\frac{\Omega_0}{2}e^{-i\omega t}\left[2\left\langle\sigma^{+}\sigma^{-}\right\rangle - 1\right] \\
\frac{d\rho_{22}}{dt} &= \frac{d}{dt}\left(1 - \left\langle\sigma^{+}\sigma^{-}\right\rangle\right) = i\frac{\Omega_0}{2}\left[e^{i\omega t}\left\langle\sigma^{+}\right\rangle - e^{-i\omega t}\left\langle\sigma^{-}\right\rangle\right]
\end{aligned} \tag{4.93}$$

Now we can easily get the time dependence of the Cartesian components by taking the appropriate linear combinations from Eqs. 4.85:

$$\begin{aligned}
\frac{d}{dt}\left\langle\sigma_x\right\rangle &= \frac{d}{dt}\left[\left\langle\sigma^{+}\right\rangle + \left\langle\sigma^{-}\right\rangle\right] = +\omega_0\left\langle\sigma_y\right\rangle - \left\langle\sigma_z\right\rangle\Omega_0\sin\omega t \\
\frac{d}{dt}\left\langle\sigma_y\right\rangle &= i\frac{d}{dt}\left[\left\langle\sigma^{-}\right\rangle - \left\langle\sigma^{+}\right\rangle\right] = -\omega_0\left\langle\sigma_x\right\rangle - \left\langle\sigma_z\right\rangle\Omega_0\cos\omega t \\
\frac{d}{dt}\left\langle\sigma_z\right\rangle &= \frac{d}{dt}\left[\left\langle 2\sigma^{+}\sigma^{-}\right\rangle - 1\right] = \Omega_0\left[\left\langle\sigma_y\right\rangle\cos\omega t + \left\langle\sigma_x\right\rangle\sin\omega t\right]
\end{aligned}$$

We can gain insight into the time-dependence of the atom density matrix by reexpressing the Hamiltonian in a coordinate frame rotating about its z axis at the same frequency and in the same propagation direction as the optical wave. The prescription that transforms the atom Hamiltonian to the rotating system is

$$\hat{H}_R = O\hat{H}O^{-} - i\hbar O\frac{\partial O^{-1}}{\partial t} \tag{4.94}$$

where the operator O is defined as

$$O = e^{i\omega t\sigma_z/2} \tag{4.95}$$

and $\hat{H}_R$ signifies the Hamiltonian in the rotating frame. It can be shown easily that the resulting form of the transformed Hamiltonian is

$$\hat{H}_{0R} = \frac{1}{2}\hbar\left(\omega_0 - \omega\right)\sigma_z = -\frac{1}{2}\hbar\Delta\omega\sigma_z$$

with the detuning $\Delta\omega = \omega - \omega_0$ and

$$\hat{V}_R = \frac{\hbar\Omega_0}{2}\left(\sigma^{-} + \sigma^{+}\right)$$

This transformation is useful because it eliminates the explicit time dependence in the Hamiltonian. Notice that as usual we have chosen the definition of the detuning so that frequencies to the red of resonance yield a negative $\Delta\omega$ while blue detuning results in a positive $\Delta\omega$. Now we can rewrite Eqs. 4.93 as

$$
\begin{aligned}
\frac{d\rho_{11}^R}{dt} &= \frac{d}{dt}\left\langle\sigma^+\sigma^-\right\rangle = i\frac{\Omega_0}{2}\left[\left\langle\sigma^-\right\rangle - \left\langle\sigma^+\right\rangle\right] \\
\frac{d\rho_{12}^R}{dt} &= \frac{d}{dt}\left\langle\sigma^-\right\rangle = -i\Delta\omega\left\langle\sigma^-\right\rangle + i\frac{\Omega_0}{2}\left[2\left\langle\sigma^+\sigma^-\right\rangle - 1\right] \\
\frac{d\rho_{21}^R}{dt} &= \frac{d}{dt}\left\langle\sigma^+\right\rangle = +i\Delta\omega\left\langle\sigma^+\right\rangle - i\frac{\Omega_0}{2}\left[2\left\langle\sigma^+\sigma^-\right\rangle - 1\right] \\
\frac{d\rho_{22}^R}{dt} &= \frac{d}{dt}\left\langle\sigma^-\sigma^+\right\rangle = i\frac{\Omega_0}{2}\left[\left\langle\sigma^+\right\rangle - \left\langle\sigma^-\right\rangle\right]
\end{aligned}
\tag{4.96}
$$

and the Cartesian components of the spin density matrices as

$$
\begin{aligned}
\frac{d}{dt}\left\langle\sigma_x^R\right\rangle &= \frac{d}{dt}\left[\left\langle\sigma^+\right\rangle + \left\langle\sigma^-\right\rangle\right] = -\Delta\omega\left\langle\sigma_y\right\rangle \\
\frac{d}{dt}\left\langle\sigma_y^R\right\rangle &= i\frac{d}{dt}\left[\left\langle\sigma^-\right\rangle - \left\langle\sigma^+\right\rangle\right] = +\Delta\omega\left\langle\sigma_x\right\rangle - \Omega_0\left\langle\sigma_z\right\rangle \\
\frac{d}{dt}\left\langle\sigma_z^R\right\rangle &= \frac{d}{dt}\left[\left\langle 2\sigma^+\sigma^-\right\rangle - 1\right] = \Omega_0\left\langle\sigma_y\right\rangle
\end{aligned}
$$

where the superscript R indicates expressions in the rotating frame.

Finally we can write the set of optical Bloch equations, Eq. 4.96, in the rotating frame in terms of a matrix as we did in Eq. 3.50

$$
\begin{bmatrix} \frac{d\rho_{11}}{dt} \\ \frac{d\rho_{12}}{dt} \\ \frac{d\rho_{21}}{dt} \\ \frac{d\rho_{22}}{dt} \end{bmatrix} = i \begin{bmatrix} 0 & \frac{\Omega_0}{2} & -\frac{\Omega_0}{2} & 0 \\ \frac{\Omega_0}{2} & -\Delta\omega & 0 & -\frac{\Omega_0}{2} \\ -\frac{\Omega_0}{2} & 0 & \Delta\omega & \frac{\Omega_0}{2} \\ 0 & -\frac{\Omega_0}{2} & \frac{\Omega_0}{2} & 0 \end{bmatrix} \begin{bmatrix} \left\langle\sigma^+\sigma^-\right\rangle \\ \left\langle\sigma^-\right\rangle \\ \left\langle\sigma^-+\right\rangle \\ \left\langle\sigma^-\sigma^+\right\rangle \end{bmatrix}
$$

and recast them in terms of a Bloch vector precessing about a torque vector, as we did previously (Eq. 3.51)

$$
\frac{d\boldsymbol{\beta}}{dt} = -\boldsymbol{\beta}\times\boldsymbol{\Omega}
$$

with

$$
\boldsymbol{\beta} = \hat{\imath}\left[\left\langle\sigma^+\right\rangle + \left\langle\sigma^-\right\rangle\right] + \hat{\jmath}\, i\left[\left\langle\sigma^+\right\rangle - \left\langle\sigma^-\right\rangle\right] + \hat{\mathbf{k}}\left[\left\langle\sigma^-\sigma+\right\rangle - \left\langle\sigma^+\sigma^-\right\rangle\right]
$$

and

$$
\boldsymbol{\Omega} = \hat{\imath}\,\Omega_0 + \hat{\jmath}\,(0) - \hat{\mathbf{k}}\Delta\omega
$$

so that

$$\begin{aligned}\frac{d\boldsymbol{\beta}}{dt} = & \hat{\boldsymbol{\imath}}\left[i\Delta\omega\left(\langle\sigma^+\rangle - \langle\sigma^-\rangle\right)\right] \\ & -\hat{\boldsymbol{\jmath}}\left[\Delta\omega\left(\langle\sigma^+\rangle + \langle\sigma^-\rangle\right) + \Omega_0\left(\langle\sigma^-\sigma^+\rangle - \langle\sigma^+\sigma^-\rangle\right)\right] \\ & +\hat{\mathbf{k}}\left[i\Omega_0\left(\langle\sigma^+\rangle - \langle\sigma^-\rangle\right)\right]\end{aligned}$$

or

$$\frac{d\boldsymbol{\beta}}{dt} = -\hat{\boldsymbol{\imath}}\left[\Delta\omega\langle\sigma_y\rangle\right] - \hat{\boldsymbol{\jmath}}\left[\Delta\omega\langle\sigma_x\rangle - \Omega_0\langle\sigma_z\rangle\right] - \hat{\mathbf{k}}\left[\Omega_0\langle\sigma_y\rangle\right] \tag{4.97}$$

We see again that in the rotating frame, the on-resonance, ($\Delta\omega = 0$), Bloch vector $\boldsymbol{\beta}$ precesses in the $\hat{\boldsymbol{\jmath}}$-$\hat{\mathbf{k}}$ plane around the torque vector $\boldsymbol{\Omega}$ pointed along the $\hat{\boldsymbol{\imath}}$ axis.

4.D Pauli Spin Matrices and Magnetic - Dipole Coupling

In this appendix we discuss another example of how the Pauli spin matrices can be used as the underlying structure to describe the physics of coupling and time evolution in a two-level system. The procedure to be followed parallels the electric dipole case. We first set up the Hamiltonian in the laboratory frame, then in the frame rotating at the Larmor frequency ω_0. From the Hamiltonian in the rotating frame and the Schrödinger equation we can express the probability of transition from the initial, ground spin state $|\beta\rangle$ to the final, excited spin state $|\alpha\rangle$.

Analogous to the interaction energy of an electric charge dipole $e\mathbf{r}$ with an electric field $\mathbf{E}$

$$w = -\boldsymbol{\mu}\cdot\mathbf{E} \tag{4.98}$$

the interaction energy of a magnetic dipole with a magnetic field is[4]

$$k = -\mathbf{m}\cdot\mathbf{B} \tag{4.99}$$

and the magnetic moment is written in terms of the Bohr magneton μ_B and the Pauli spin operator $\boldsymbol{\sigma}$

$$\mathbf{m} = -\frac{e\hbar}{2m_e}\boldsymbol{\sigma} = -\mu_B\boldsymbol{\sigma} = -\frac{1}{2}\gamma\hbar\boldsymbol{\sigma} \tag{4.100}$$

The constant γ is called the *gyromagnetic ratio*; and, because of the choice of negative sign for the electron charge, the magnetic dipole direction must be defined opposite to the electron angular momentum direction. Analogous to Eq. 1.38 relating the displacement field $\mathbf{D}$ to the electric field $\mathbf{E}$ and polarization

[4]Note that Eqs. 4.98 and 4.99 involve permanent dipoles. The interaction energy involving *induced* dipoles contains an extra factor of $\frac{1}{2}$ (see Eq. 4.38).

$\mathbf{P}$, there exists a relation between the magnetic field $\mathbf{H}$, the magnetic induction field $\mathbf{B}$ and the magnetization $\mathbf{M}$,

$$\mathbf{H} = \frac{\mathbf{1}}{\mu_0}\mathbf{B} - \mathbf{M} \tag{4.101}$$

Just as we write the polarization as an ensemble density of the average electric transition dipole,

$$\mathbf{P} = \frac{N}{V}\langle\boldsymbol{\mu}\rangle \tag{4.102}$$

so we can write the magnetization as an ensemble density of the average magnetic dipole

$$\mathbf{M} = \frac{N}{V}\langle\mathbf{m}\rangle = -\frac{\gamma\hbar}{2}\frac{N}{V}\langle\boldsymbol{\sigma}\rangle \tag{4.103}$$

Furthermore, just as we have for the energy of interaction W between $\mathbf{P}$ and $\mathbf{E}$

$$W = -\mathbf{P}\cdot\mathbf{E}$$

so we have the energy of interaction K between $\mathbf{M}$ and $\mathbf{B}$

$$K = -\mathbf{M}\cdot\mathbf{B} \tag{4.104}$$

Passing to quantum mechanics, the Hamiltonian operator representing this interaction energy, in vacuum, is clearly

$$\hat{H}_0 = \frac{1}{2}\gamma\hbar B_0\sigma_z$$

and the Schrödinger equation for the two spin states $|\alpha\rangle$ and $|\beta\rangle$ becomes

$$\hat{H}_0|\beta\rangle = \frac{1}{2}\gamma\hbar B_0|\beta\rangle \tag{4.105}$$

$$\hat{H}_0|\alpha\rangle = -\frac{1}{2}\gamma\hbar B_0|\alpha\rangle \tag{4.106}$$

with the energy difference

$$\Delta E = \gamma\hbar B_0 = \hbar\omega_0 \tag{4.107}$$

Now we couple these two states with the classical oscillating magnetic field, circularly polarized in the $x-y$ plane and propagating in the z direction:

$$\mathbf{b_1} = b_1(\hat{\boldsymbol{\imath}}\cos\omega t + \hat{\boldsymbol{\jmath}}\sin\omega t) \tag{4.108}$$

The Hamiltonian becomes

$$\hat{H} = \frac{1}{2}\gamma\hbar\left[B_0\sigma_z + b_1\left(\sigma_x\cos\omega t + \sigma_y\sin\omega t\right)\right] \tag{4.109}$$

$$= \frac{1}{2}\gamma\hbar\left[B_0\sigma_z + b_1\left(e^{i\omega t}\sigma^+ + e^{-i\omega t}\sigma^-\right)\right] \tag{4.110}$$

and if we change to the coordinate system rotating at ω by using the same transformation operator employed in Eqs. 4.94 and 4.95, the Hamiltonian becomes

$$\hat{H}_R = \frac{1}{2}\gamma\hbar\left[\left(B_0 - \frac{\omega}{\gamma}\right)\sigma_z + b_1\left(\sigma^+ + \sigma^-\right)\right] \tag{4.111}$$

$$= \frac{1}{2}\hbar\left(\omega_0 - \omega\right)\sigma_z + \frac{1}{2}\gamma\hbar b_1\sigma_x \tag{4.112}$$

$$= -\frac{1}{2}\hbar\Delta\omega_B\sigma_z + \frac{1}{2}\gamma\hbar b_1\sigma_x \tag{4.113}$$

In the last line we have defined the "detuning" of the frequency ω of b_1 from the precession frequency ω_0 of a magnetic moment about the constant magnetic field $\mathbf{B_0}$ as $\Delta\omega_B$. Now we seek the probability of finding the system in the excited state at some time t in the representation spanned by the two basis states $|\alpha\rangle$ and $|\beta\rangle$

$$|\psi(t)\rangle = c_\alpha(t)\,|\alpha\rangle + c_\beta(t)\,|\beta\rangle$$

by solving the Schrödinger equation

$$\left(\hat{H}_R - E\right)|\psi(t)\rangle = 0$$

subject to the initial condition that, at the time t_0 when the coupling field b_1 is switched on, the system occupies the ground spin state ($c_\beta(t_0) = 0$ and $c_\alpha(t_0) = 1$). The probability of finding the system in the spin state $|\beta\rangle$ at time t is

$$\left|c_\beta(t)\right|^2 = \frac{\left(\gamma b_1\right)^2}{\left(\omega - \gamma B_0\right)^2 + \left(\gamma b_1\right)^2}\sin^2\left[\frac{\left[\left(\omega - \gamma B_0\right)^2 + \left(\gamma b_1\right)^2\right]^{1/2}}{2}t\right] \tag{4.114}$$

We define a spin Rabi frequency Ω_B analogous to the optical Rabi frequency Ω

$$\Omega_B^2 = \left(\omega - \gamma B_0\right)^2 + \left(\gamma b_1\right)^2 \tag{4.115}$$

so

$$\left|c_\beta(t)\right|^2 = \rho_{\beta\beta} = \frac{\left(\gamma b_1\right)^2}{\Omega_B^2}\sin^2\frac{\Omega_B}{2}t \tag{4.116}$$

When the oscillating magnetic field is tuned to the resonance frequency ω_0, we have

$$\left|c_\beta(t)\right|^2 = \rho_{\beta\beta} = \sin^2\frac{\left(\gamma b_1\right)t}{2} = \sin^2\frac{\Omega_B^0}{2}t$$

These results are analogous to the solutions of the optical Bloch equations expressed in Eq. 4.6 :

$$\rho_{22} = \frac{\left|\Omega_0\right|^2}{\Omega^2}\sin^2\left(\frac{\Omega}{2}t\right)$$

4.E Time Evolution of the Magnetic Dipole - Coupled Atom Density Matrix

Once again, just as in the case of the electric dipole moment, the equation of motion of the density matrix is given by the *Liouville equation*:

$$\frac{d\rho(t)}{dt} = \frac{i}{\hbar}\left[\rho(t), \hat{H}\right] \tag{4.117}$$

We switch to the rotating frame by invoking the transformation in Eqs. and 4.94, 4.95 and write the time dependence of the density matrix elements in the rotating frame with the help of Eq. 3.43. In the rotating frame the coupling matrix elements become

$$V_{\alpha\beta} = V_{\beta\alpha} = \frac{1}{2}\gamma\hbar b_1$$

and we have

$$\frac{d\rho^R_{\alpha\alpha}}{dt} = \frac{1}{2}\gamma b_1 \langle\sigma_y\rangle \tag{4.118}$$

$$\frac{d\rho^R_{\alpha\beta}}{dt} = -i\frac{\Delta\omega_B}{2}\left[\langle\sigma_x\rangle - i\langle\sigma_y\rangle\right] + i\frac{\gamma b_1}{2}\langle\sigma_z\rangle \tag{4.119}$$

$$\frac{d\rho^R_{\beta\alpha}}{dt} = i\frac{\Delta\omega_B}{2}\left[\langle\sigma_x\rangle + i\langle\sigma_y\rangle\right] - i\frac{\gamma b_1}{2}\langle\sigma_z\rangle \tag{4.120}$$

$$\frac{d\rho^R_{\beta\beta}}{dt} = -\frac{1}{2}\gamma b_1 \langle\sigma_y\rangle \tag{4.121}$$

Note that here we have the expressed time dependence of the density matrix using Cartesian matrix elements while in Eq. 4.96 we used the circular matrix elements. Now we can find the time dependencies of the matrix elements of the Cartesian components of σ by taking appropriate linear combinations of Eqs. 4.118 – 4.121:

$$\begin{aligned}
\frac{d\langle\sigma_x\rangle}{dt} &= \frac{d}{dt}\left[(\rho_{\alpha\beta} + \rho_{\beta\alpha})\right] = -\Delta\omega_B\langle\sigma_y\rangle \\
\frac{d\langle\sigma_y\rangle}{dt} &= i\frac{d}{dt}\left[(\rho_{\alpha\beta} - \rho_{\beta\alpha})\right] = \Delta\omega_B\langle\sigma_x\rangle - \gamma b_1\langle\sigma_z\rangle \\
\frac{d\langle\sigma_z\rangle}{dt} &= \frac{d}{dt}(\rho_{\alpha\alpha} - \rho_{\beta\beta}) = \gamma b_1\langle\sigma_y\rangle
\end{aligned}$$

(4.122)

The time evolution of the quantum two-level magnetic spin system (Eqs. 4.122) is very similar to the classical spin precession. The similarity is hardly surprising since any quantum semiclassical theory uses only classical fields. It does mean, however, that the precessing vector model carries over from the classical to the quantum description of a magnetic moment interacting with a strong constant magnetic field B_0 and a rotating magnetic field b_1.

Finally we can once again write the set of optical Bloch equations, Eq. 4.96, in the rotating frame in terms of a matrix as we did in Eq. 3.50

$$\frac{d\boldsymbol{\beta}}{dt} = -\boldsymbol{\beta} \times \boldsymbol{\Omega}$$

with

$$\boldsymbol{\beta} = \hat{\imath}\langle\sigma_x\rangle - \hat{\jmath}\langle\sigma_y\rangle - \mathbf{k}\langle\sigma_z\rangle$$

and

$$\boldsymbol{\Omega} = \hat{\imath}\gamma b_1 - \hat{\mathbf{k}}\Delta\omega_B$$

so that

$$\frac{d\boldsymbol{\beta}}{dt} = \hat{\imath}\left[-\Delta\omega_B\langle\sigma_y\rangle\right] - \hat{\jmath}\left[\Delta\omega_B\langle\sigma_x\rangle - \gamma b_1\langle\sigma_z\rangle\right] + \hat{\mathbf{k}}\left[-\gamma b_1\langle\sigma_z\rangle\right]$$

The motion of the Bloch vector in two-level magnetic dipole coupling corresponds to the motion of the Bloch vector in two-level electric dipole coupling as is evident from Eq. 4.97. The Bloch vector precesses about $\boldsymbol{\Omega}$, which, with γb_1 tuned to the Larmor precession frequency, points in the $\hat{\imath}$ direction

Chapter 5

Quantized Fields and Dressed States

5.1 Introduction

So far we have only expressed the optical field as a classical standing or traveling wave while regarding our two-level atom as a quantum - mechanical entity subject to the time-dependent, oscillatory perturbation of the wave. This approach leads quite naturally to populations and coherences oscillating among the states of the atom. However, for strong-field problems involving a significantly modified atomic energy spectrum, a nonperturbative, time-independent approach can be fruitful. Time-independent solutions to the atom–field Schrödinger equation are called *dressed states.* They were first used to interpret the "doubling" of molecular rotation spectra in the presence of intense, classical RF fields. The semiclassical approach is adequate for a wide variety of phenomena and has the virtue of mathematical simplicity and familiarity. However, sometimes it is worthwhile to consider the field as a quantum-mechanical entity as well, and the atom–field interaction then becomes an exchange of field quanta (photons) with the atom. This approach leads us to express the photon-number states and the discrete states of the atom on an equal footing and to write the state functions of the atom-plus-field system in a basis of product photon and atom states. Diagonalization of the dipole coupling terms in the system Hamiltonian between the photon–atom states also gives rise to time-independent, dressed-state solutions of the full quantal Schrödinger equation.

Since the photon–atom product state basis is usually encountered in contemporary research literature, we will begin this chapter with the development of the quantized light field and then express the atom–field interaction in fully quantized form. Then we will examine some illustrative examples of how the dressed-state picture can provide useful insights to light–matter interactions. In Appendix 5.A we will show how semiclassical dressed states can also be obtained.

5.2 Classical Fields and Potentials

The essential idea behind field quantization is the substitution of a set of quantum-mechanical harmonic oscillators for the classical oscillators discussed in Section 1.1. In order to carry out this quantization in the simplest way, however, we introduce two new quantities: the scalar potential ϕ and the vector potential $\mathbf{A}$. The conventional starting point is Maxwell's equations, which we write as

$$\nabla \times \mathbf{E} = -\frac{\partial \mathbf{B}}{\partial t}$$
$$\nabla \times \mathbf{B} = \frac{1}{c^2}\frac{\partial \mathbf{E}}{\partial t} + \mu_0 \mathbf{J}$$
$$\nabla \cdot \mathbf{E} = \frac{\sigma}{\epsilon_0}$$
$$\nabla \cdot \mathbf{B} = 0$$

where $\mathbf{J}$ is the current density and σ is the charge density. The vector potential is related to the magnetic and electric fields by two key equations. The vector potential is defined in terms of the magnetic field by

$$\mathbf{B} = \nabla \times \mathbf{A} \tag{5.1}$$

and is related to the scalar potential and the electric field by

$$\mathbf{E} = -\nabla\phi - \frac{\partial \mathbf{A}}{\partial t} \tag{5.2}$$

Now it is a standard result from electromagnetic theory that $\mathbf{A}$ and ϕ can be specified in different forms while leaving the physically observable fields $\mathbf{E}$ and $\mathbf{B}$ invariant. These forms or *gauges* are related by what are called *gauge transformations.* One particularly useful gauge is defined such that

$$\nabla \cdot \mathbf{A} = 0$$

This condition puts the electromagnetic field into the *Coulomb gauge.* With the choice of the Coulomb gauge the second and third Maxwell equations can be expressed as

$$-\nabla^2\mathbf{A} + \frac{1}{c^2}\left[\frac{\partial^2 \mathbf{A}}{\partial t^2} + \frac{\partial}{\partial t}\nabla\phi\right] = \mu_0 \mathbf{J} \tag{5.3}$$

and

$$-\nabla^2\phi = \frac{\sigma}{\epsilon_0} \tag{5.4}$$

These two equations determine the vector and scalar potentials if the actual current and charge density distributions of the problem are specified. Equation 5.4 is particularly simple since it involves *only* the scalar potential field, and the formal solution is the familiar Poisson equation:

$$\phi(\mathbf{r}) = \frac{1}{4\pi\epsilon_0}\int \frac{\sigma(r')}{|\mathbf{r} - \mathbf{r}'|}d^3r'$$

In order to obtain an equation involving *only* the vector field and the current density, we use Helmholtz's theorem to write the current density as the sum of transverse and longitudinal components

$$\mathbf{J} = \mathbf{J}_T + \mathbf{J}_L$$

where the terms transverse and longitudinal are defined by the following two conditions,

$$\nabla \cdot \mathbf{J}_T = 0$$
$$\nabla \times \mathbf{J}_L = 0$$

Then it can be shown that the longitudinal component of $\mathbf{J}$ is associated entirely with the scalar potential

$$\mathbf{J}_L = \varepsilon_0 \frac{\partial}{\partial t} \nabla \phi$$

and therefore from Eq. 5.3 we have

$$-\nabla^2 \mathbf{A} + \frac{1}{c^2} \frac{\partial^2 \mathbf{A}}{\partial t^2} = \mu_0 \mathbf{J}_T \tag{5.5}$$

which shows that the transverse component of $\mathbf{J}$ is associated only with the vector potential. In free space where there are no currents, Eq. 5.5 becomes

$$-\nabla^2 \mathbf{A} + \frac{1}{c^2} \frac{\partial^2 \mathbf{A}}{\partial t^2} = 0 \tag{5.6}$$

and we seek plane-wave solutions to this equation in the form

$$\mathbf{A} = \sum_k \left\{ \mathbf{A}_\mathbf{k}(t) e^{i\mathbf{k}\cdot\mathbf{r}} + \mathbf{A}_\mathbf{k}^*(t) e^{-i\mathbf{k}\cdot\mathbf{r}} \right\}$$

Now we subject these plane-wave components to periodic boundary conditions corresponding to the cavity boundary conditions of Section 1.2

$$k_m = \frac{2\pi\nu_m}{L}; \; m = x, y, z$$

where, as earlier, $V = L^3$ is the cavity volume. Note that each A_k and A_k^* must satisfy Eq. 5.6 independently and $\nabla^2 A_k = -k^2 A_k$. Then we can write Eq. 5.6 for each A_k component as

$$k^2 \mathbf{A}_\mathbf{k}(t) + \frac{1}{c^2} \frac{\partial^2 \mathbf{A}_\mathbf{k}(t)}{\partial t^2} = 0$$

or, with $\omega_k = ck$, as

$$\omega_k^2 \mathbf{A}_\mathbf{k}(t) + \frac{\partial^2 \mathbf{A}_\mathbf{k}(t)}{\partial t^2} = 0 \tag{5.7}$$

The same equation obtains for $\mathbf{A}_\mathbf{k}^*$. It is obvious from Eq. 5.7 that the free-space time dependence of the vector potential is just an oscillatory factor with frequency ω_k:

$$\mathbf{A}_\mathbf{k}(t) = \mathbf{A}_\mathbf{k} e^{i(\mathbf{k}\cdot\mathbf{r} - \omega_k t)} \tag{5.8}$$

The factor $\mathbf{A_k}$ represents the amplitude and polarization direction of the vector potential wave. The amplitude is complex and can be written in terms of a real part and imaginary part. We choose the form for these parts by introducing the generalized momentum and position coordinates for the Hamiltonian of the classical oscillator for the kth mode:

$$\mathbf{A_k} = \frac{1}{\sqrt{4\epsilon_0 V \omega_k^2}} \left(\omega_k Q_k + iP_k \right) \hat{\varepsilon}_k \tag{5.9}$$

Note that P_k and Q_k are scalars and the vector property of $\mathbf{A_k}$ comes from the polarization unit vector $\hat{\varepsilon}_k$. As we saw when we were counting cavity modes in Section 1.1.2, there are two independent polarization directions per mode. Now we are going to express the energy of the kth mode in terms of $\mathbf{A_k}(t)$ and $\mathbf{A_k^*}(t)$. We remember from Eq. 1.7 that the total period-averaged energy of the electromagnetic field can be written in terms of the electric field as

$$\bar{U} = \frac{1}{2} \int \epsilon_0 \left| \mathbf{E} \right|^2 dV$$

and this total energy is the sum of the energies of the component modes. Therefore we can write for each component

$$\bar{U_k} = \frac{1}{2} \int \epsilon_0 \left| \mathbf{E}_k \right|^2 dV \tag{5.10}$$

Now for each kth component, from the definition of the vector potential in terms of the electric field and the scalar potential, Eq. 5.2, and remembering that there are no electric charges in the cavity ($\phi = 0$), we can write

$$\mathbf{E}_k = i\omega_k \left\{ \mathbf{A_k} e^{i(\mathbf{k}\cdot\mathbf{r} - \omega_k t)} - \mathbf{A_k^*} e^{-i(\mathbf{k}\cdot\mathbf{r} - \omega_k t)} \right\} \tag{5.11}$$

and that therefore the period-averaged field energy is

$$\bar{U_k} = \frac{1}{2} \int \epsilon_0 \left| \mathbf{E}_k \right|^2 dV = 2\epsilon_0 V \omega_k^2 \mathbf{A_k} \cdot \mathbf{A_k^*} \tag{5.12}$$

The final step is to substitute the transformation Eq. 5.9 for $\mathbf{A_k}$ and $\mathbf{A_k^*}$ in Eq. 5.12. The result is

$$\bar{U_k} = \frac{1}{2} \left(P_k^2 + \omega_k^2 Q_k^2 \right) \tag{5.13}$$

which, of course, is the standard form for the one-dimensional classical harmonic oscillator. The terms P_k and Q_k are the canonically conjugate "momentum" and "position" variables of the classical harmonic oscillator Hamiltonian. In terms of the ordinary momentum $p = mv$ and position q, the Hamiltonian for each independent polarization direction is

$$\bar{U_k} = \frac{1}{2} \left(\frac{p_k^2}{2m} + \frac{m\omega_k^2}{2} q_k^2 \right)$$

with $\omega = \sqrt{\frac{k}{m}}$, where k is the oscillator force constant. The variables are simply mass-weighted by

$$q \longrightarrow \sqrt{\frac{2}{m}}Q \text{ and } p \longrightarrow \sqrt{2m}P$$

but since there is no mass associated with modes of the electromagnetic field, the expression of the energy in terms of the more abstract P and Q is more appropriate. The total energy for the cavity is the sum over the k modes and the two independent polarization directions $\hat{\varepsilon_k}$, :

$$\bar{U} = 2\sum_k \bar{U_k} = \sum_k P_k^2 + \omega_k^2 Q_k^2$$

It should be noted that the cavity mode components of the magnetic field can also be constructed from the vector potential components using Eq. 5.1 such that

$$\mathbf{B}_k = i\mathbf{k} \times \left\{ \mathbf{A}_k e^{i(\mathbf{k}\cdot\mathbf{r}-i\omega_k t} - \mathbf{A}_{\mathbf{k}}^* e^{-i(\mathbf{k}\cdot\mathbf{r}-\omega_k t)} \right\} \tag{5.14}$$

5.3 Quantized Oscillator

Now our task is to transform the classical expression, Eq. 5.13, to its quantum-mechanical counterpart. In order to carry out this transformation in the most convenient way, we have to invest some time in the development of an operator algebra involving the operators corresponding to the classical P and Q of Eq. 5.13. We use the usual correspondence principal to go from variables to operators in order to form the quantum-mechanical Hamiltonian of the one-dimensional oscillator

$$P \longrightarrow \hat{p} \text{ and } Q \longrightarrow \hat{q}$$

which is then given by

$$\hat{H} = \frac{1}{2}\left(\hat{p}^2 + \omega^2 \hat{q}^2\right)$$

where $\hat{q}$ and $\hat{p}$ are the conjugate position and momentum operators, respectively. They obey the usual commutator relation conjugate variables

$$[\hat{q}, \hat{p}] = i\hbar$$

Now we define two new operators that are linear combinations of $\hat{p}$ and $\hat{q}$, :

$$\hat{a} = \frac{1}{\sqrt{2\hbar\omega}}\left(\omega\hat{q} + i\hat{p}\right) \tag{5.15}$$

$$\hat{a}^\dagger = \frac{1}{\sqrt{2\hbar\omega}}\left(\omega\hat{q} - i\hat{p}\right)$$

These operators are called the *annihilation operator* ($\hat{a}$) and the *creation operator* ($\hat{a}^\dagger$) for reasons that will become evident shortly. From these definitions it is easy to show that

$$\begin{aligned}\hat{a}^\dagger\hat{a} &= \frac{1}{2\hbar\omega}(\hat{p}^2 + \omega^2\hat{q}^2 + i\omega\hat{q}\hat{p} - i\omega\hat{p}\hat{q}) \\ &= \frac{1}{\hbar\omega}\left(\hat{H} - \frac{1}{2}\hbar\omega\right)\end{aligned} \tag{5.16}$$

and

$$\hat{a}\hat{a}^\dagger = \frac{1}{\hbar\omega}\left(\hat{H} + \frac{1}{2}\hbar\omega\right) \tag{5.17}$$

Evidently from Eqs. 5.16 and 5.17 the commutator relation for annihilation and creation operators is

$$\left[\hat{a}, \hat{a}^\dagger\right] = \hat{a}\hat{a}^\dagger - \hat{a}^\dagger\hat{a} = 1 \tag{5.18}$$

and the oscillator Hamiltonian can be expressed in terms of the product of creation and annihilation operators as

$$\hat{H} = \hbar\omega\left(\hat{a}^\dagger\hat{a} + \frac{1}{2}\right) = \hbar\omega\left(\hat{n} + \frac{1}{2}\right) \tag{5.19}$$

where we have defined the *number operator* $\hat{n}$ as

$$\hat{n} = \hat{a}^\dagger\hat{a} \tag{5.20}$$

Now we will denote the eigenstates of the harmonic oscillator $|n\rangle$ so that

$$\hat{H}\,|n\rangle = \hbar\omega\left(\hat{a}^\dagger\hat{a} + \frac{1}{2}\right)|n\rangle = E_n\,|n\rangle \tag{5.21}$$

and investigate the effect of the $\hat{a}$ and $\hat{a}^\dagger$ operators on $|n\rangle$. First we multiply Eq. 5.21 from the left by $\hat{a}$, which gives

$$\hbar\omega\left(\hat{a}\hat{a}^\dagger\hat{a} + \frac{\hat{a}}{2}\right)|n\rangle = E_n\hat{a}\,|n\rangle$$

and then substitute $\hat{a}\hat{a}^\dagger = 1 + \hat{a}^\dagger\hat{a}$ from the commutation relation, Eq. 5.18. The result is

$$\hbar\omega\left[\left(1 + \hat{a}^\dagger\hat{a} + \frac{1}{2}\right)\hat{a}\,|n\rangle = E_n\hat{a}\,|n\rangle\right]$$

which, from Eq. 5.19, can be written as

$$\hat{H}\,\hat{a}\,|n\rangle = (E_n - \hbar\omega)\,\hat{a}\,|n\rangle$$

Clearly $\hat{a}\,|n\rangle$ is an eigenstate of the oscillator Hamiltonian with eigenvalue $E_n - \hbar\omega$. So the effect of the annihilation operator on $|n\rangle$ is to transform it to an new eigenstate with energy lower by an amount $\hbar\omega$. One might say that $\hat{a}$ has annihilated a quantum of energy $\hbar\omega$ in the quantized oscillator. The new eigenstate is denoted

$$\hat{a}\,|n\rangle = |n-1\rangle$$

with eigenvalue

$$E_n - \hbar\omega = E_{n-1}$$

Similar reasoning, but starting with left multiplication by $\hat{a}^\dagger$ of Eq. 5.21, leads to the anticipated effect of $\hat{a}^\dagger$ on $|n\rangle$:

$$\hat{H}\,\hat{a}^\dagger\,|n\rangle = (E_n + \hbar\omega)\,\hat{a}^\dagger\,|n\rangle$$

We see that $\hat{a}^\dagger$ operating on $|n\rangle$ creates a new eigenstate of the oscillator Hamiltonian whose eigenenergy is increased by a quantum $\hbar\omega$. The corresponding notations are

$$\hat{a}^\dagger\,|n\rangle = |n+1\rangle$$

and

$$E_n + \hbar\omega = E_{n+1} \tag{5.22}$$

Of course, the quantized oscillator states are orthogonal, and if we impose the usual normalization conditions

$$\langle n|n\rangle = 1$$

we find the following results:

$$\hat{a}\,|n\rangle = \sqrt{n}\,|n-1\rangle$$
$$\hat{a}^\dagger\,|n\rangle = \sqrt{(n+1)}\,|n+1\rangle$$

Having established the normalization constants, we can appreciate why $\hat{n}$ is called the *number operator*. Notice the effect of the number operator on $|n\rangle$. From Eq. 5.20, we obtain

$$\hat{n}\,|n\rangle = \hat{a}^\dagger\hat{a}\,|n\rangle = n\,|n\rangle \tag{5.23}$$

We see that the oscillator states $|n\rangle$ are eigenstates of $\hat{n}$ with *eigenvalues equal to the number of energy quanta in the state above the zero point.*

Repeated applications of $\hat{a}$ to the eigenstates of the oscillator lower the energy in steps of $\hbar\omega$ until the energy reaches the zero point. Thus there will be a state such that

$$\hat{H}\,\hat{a}\,|1\rangle = (E_1 - \hbar\omega)\,\hat{a}\,|1\rangle = \hat{H}\,|0\rangle = E_0\,|0\rangle$$

But from Eq. 5.21

$$\hat{H}\,|0\rangle = \hbar\omega\left(\hat{a}^\dagger\hat{a} + \frac{1}{2}\right)|0\rangle = E_0\,|0\rangle$$

and taking into account from Eq. 5.23 that

$$\hat{a}^\dagger\hat{a}\,|0\rangle = 0$$

We see that

$$\hat{H}\,|0\rangle = \frac{\hbar\omega}{2}\,|0\rangle = E_0\,|0\rangle$$

so evidently the zero-point energy is

$$E_0 = \frac{\hbar\omega}{2}$$

It is now clear that the set of eigenvalues of the one dimensional quantized harmonic oscillator consists of a ladder of energies equally spaced by $\hbar\omega$, the bottom rung of which is positioned at the zero-point energy,

$$E_n = \left(n + \frac{1}{2}\right)\hbar\omega; \qquad n = 0, 1, 2, 3, \ldots$$

5.4 Quantized Field

The quantization of the radiation field proceeds quite straightforwardly from the classical expressions for the vector potential field modes (Eqs. 5.9) in two steps. First we substitute the operators $\hat{q}_k, \hat{p}_k$ for the classical variables Q_k, P_k

$$\mathbf{A_k} = \frac{1}{\sqrt{4\epsilon_0 V\omega_k^2}}\left(\omega_k Q_k + iP_k\right)\hat{\varepsilon}_k \longrightarrow \hat{\mathbf{A}}_\mathbf{k} = \frac{1}{\sqrt{4\epsilon_0 V\omega_k^2}}\left(\omega_k\hat{q}_k + i\hat{p}_k\right)\hat{\varepsilon}_k$$

$$\mathbf{A_k}^* = \frac{1}{\sqrt{4\epsilon_0 V\omega_k^2}}\left(\omega_k Q_k - iP_k\right)\hat{\varepsilon}_k \longrightarrow \hat{\mathbf{A}}_\mathbf{k}^* = \frac{1}{\sqrt{4\epsilon_0 V\omega_k^2}}\left(\omega_k\hat{q}_k - i\hat{p}_k\right)\hat{\varepsilon}_k$$

then the expressions for the annihilation and creation operators (Eq. 5.15) :

$$\hat{\mathbf{A}}_k = \frac{1}{\sqrt{4\epsilon_0 V\omega_k^2}}\left(\omega_k\hat{q}_k + i\hat{p}_k\right)\hat{\varepsilon}_k = \sqrt{\frac{\hbar}{2\epsilon_0 V\omega_k}}\hat{a}_k\hat{\varepsilon}_k \tag{5.24}$$

$$\hat{\mathbf{A}}_k^* = \frac{1}{\sqrt{4\epsilon_0 V\omega_k^2}}\left(\omega_k\hat{q}_k - i\hat{p}_k\right)\hat{\varepsilon}_k = \sqrt{\frac{\hbar}{2\epsilon_0 V\omega_k}}\hat{a}_k^\dagger\hat{\varepsilon}_k$$

We see from Eq. 5.24 that individual cavity-mode components k of the quantized vector potential field operators bear a very simple relation to the annihilation and creation operators of that mode. From Eqs. 5.11 and 5.14 we can construct the electric and magnetic field operators for the cavity modes:

$$\hat{\mathbf{E}}_k = i\sqrt{\frac{\hbar\omega_k}{2\epsilon_0 V}}\left\{\hat{a}_k e^{i(\mathbf{k}\cdot\mathbf{r}-\omega_k t)} - \hat{a}_k^\dagger e^{-i(\mathbf{k}\cdot\mathbf{r}-\omega_k t)}\right\}\hat{\varepsilon}_k \tag{5.25}$$

$$\hat{\mathbf{B}}_k = i\sqrt{\frac{\hbar\omega_k}{2\epsilon_0 V}}\left\{\hat{a}_k e^{i(\mathbf{k}\cdot\mathbf{r}-\omega_k t)} - \hat{a}_k^\dagger e^{-i(\mathbf{k}\cdot\mathbf{r}-\omega_k t)}\right\}\mathbf{k}\times\hat{\varepsilon}_k \tag{5.26}$$

We can calculate the period-averaged energy of the kth mode in the cavity by invoking the quantized field equivalent of Eq. 5.10

$$\bar{U}_k = \frac{\epsilon_0}{2}\int \langle n|\,\hat{\mathbf{E}}_k\cdot\hat{\mathbf{E}}_k\,|n\rangle$$

which yields, when substituting Eq. 5.25 and taking into account the two orthogonal polarization directions:

$$\bar{U}_k = \hbar\omega_k\left(\frac{1}{2} + n\right) \tag{5.27}$$

The total energy of the field is just the sum over modes:

$$\bar{U} = \sum_k \hbar\omega_k(\frac{1}{2} + n)$$

This result is, of course, exactly what Planck had suggested (although strictly speaking his suggestion was the quantization of the oscillators in the walls of the conducting sphere, not the field), to account for the spectral intensity distribution radiating from a blackbody. We see now that it follows naturally from the quantization of the field modes in the cavity (see Section 1.26).

5.5 Atom–Field States

5.5.1 Second quantization

Now that we have a clear picture of the quantized field with mode energies given by Eq. 5.27 and *photon number states* given by the eigenstates of the quantized harmonic oscillator, $|n\rangle$, we are in a position to consider our two-level atom interacting with this quantized radiation field. If we exclude spontaneous emission and stimulated processes for the time being, the Hamiltonian of the combined system of atom plus field is

$$\hat{H} = \hat{H}_A + \hat{H}_F + \hat{H}_I \tag{5.28}$$

where $\hat{H}_A$ is the Hamiltonian of the atom,

$$\hat{H}_A = -\frac{\hbar\omega_0}{2}|1\rangle\langle 1| + \frac{\hbar\omega_0}{2}|2\rangle\langle 2| \tag{5.29}$$

with $|1\rangle, |2\rangle$ the lower and upper atomic states, respectively; $\hat{H}_F$ the Hamiltonian of the quantized field, expressed by Eq. 5.19, and $\hat{H}_I$ the atom–field interaction. For the noninteracting Hamiltonian, $\hat{H} = \hat{H}_A + \hat{H}_F$, the eigenstates are simply product states of the atom and the photon number states

$$N_1\,|1\rangle\,|n\rangle = N_1\,|1;n\rangle \tag{5.30}$$

or

$$N_2\,|2\rangle\,|n\rangle = N_2\,|2;n\rangle \tag{5.31}$$

where N_1, N_2 are as yet undetermined normalization constants. Figure 5.1 shows how the product eigenenergies consist of two ladders offset by the detuning energy $\hbar\Delta\omega$. We have written the atom Hamiltonian operator Eq. 5.29 as a sum over pairs of state operators, a form closely related to the definition of the density matrix operator, Eq. 3.1, but also reminiscent of the sum over creation and annihilation operators used to construct the field Hamiltonian, Eq. 5.25. In fact, we can use operator pairs to move up and down the ladder of atom levels in a way analogous to the use of creation and annihilation operators to move up and down the ladder of quantized field levels. The usefulness of this point

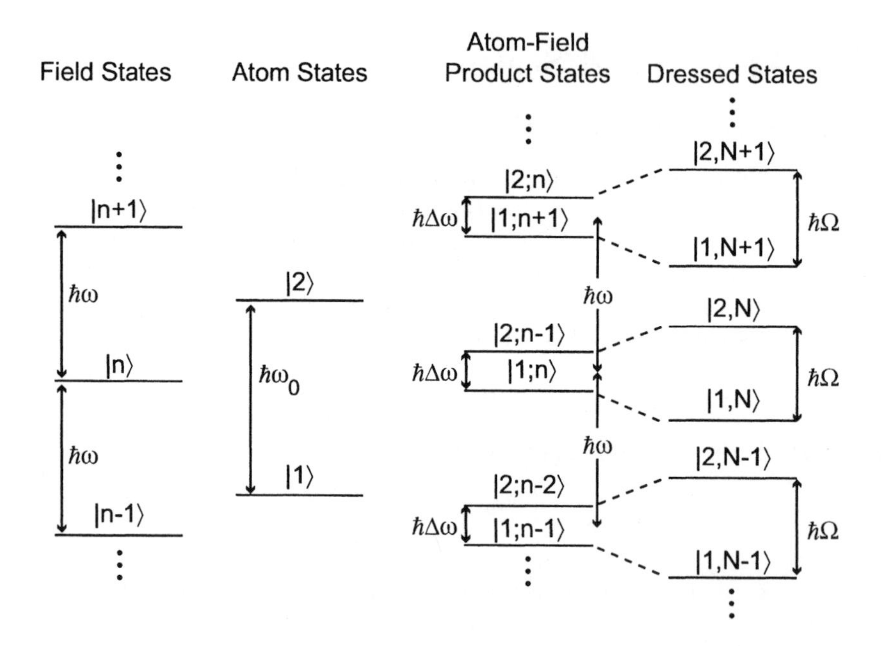

Figure 5.1: Left: photon number states and the two stationary states of the two-level atom. Middle: double ladder showing the product state basis of photon number and atom states. Right: dressed states constructed from diagonalizing the full Hamiltonian in the product state basis.

of view arises from the fact that atom–field interaction terms $\hat{H}_I$ can then also be expressed as a sum over ordered, sequential field–atom raising and lowering operators. The various terms in the sum can be easily visualized by simple diagrams that suggest a straightforward procedure for handling multilevel atoms and nonlinear atom–field interactions. The expression of the atom Hamiltonian in this form is called *second quantization,* a discussion of which is worth a brief digression.

We assume that we have already solved the Schrödinger equation for the atom and that we know all the eigenfunctions and eigenenergies. Therefore for any state we can write

$$\hat{H}_A \left| j \right\rangle = \hbar\omega_j \left| j \right\rangle$$

That accomplishment is referred to as *first quantization.* We can now write "unity" formally as a closure relation on this complete set of atom eigenfunctions

$$\sum_i \left| i \right\rangle \left\langle i \right| = 1$$

and then write $\hat{H}_A$ by inserting it between two expressions of "unity"

$$\hat{H}_A = \sum_i \left| i \right\rangle \left\langle i \right| \hat{H}_A \sum_j \left| j \right\rangle \left\langle j \right| = \hbar\omega_j \left| j \right\rangle \left\langle j \right| \tag{5.32}$$

Now we take the dipole operator defined in Eqs. 2.7, and 2.8 and surround it with closure sums in the same way:

$$\boldsymbol{\mu} = \sum_i \left| i \right\rangle \left\langle i \right| \boldsymbol{\mu} \sum_j \left| j \right\rangle \left\langle j \right| = \sum_{ij} \mu_{ij} \left| i \right\rangle \left\langle j \right| \tag{5.33}$$

Note that $\boldsymbol{\mu}$ can have (in fact will have) only off-diagonal elements. Now we use Eqs. 5.25 and 5.33 together in the atom–field interaction Hamiltonian $\hat{H}_I = \boldsymbol{\mu}\cdot\mathbf{E}$

$$\hat{H}_I = i\sum_k \sum_{ij} \left(\frac{\hbar\omega_k}{2\varepsilon_0 V}\right)^{1/2} \mu_{ij}\cdot\varepsilon_k \left\{\hat{a}_k e^{i(\mathbf{k}\cdot\mathbf{r}-\omega_k t)} - \hat{a}_k^\dagger e^{-i(\mathbf{k}\cdot\mathbf{r}-\omega_k t)}\right\} \left| i \right\rangle \left\langle j \right|$$

and for our two-level atom interacting with a single-mode field we just have

$$\hat{H}_I = i\left(\frac{\hbar\omega_k}{2\varepsilon_0 V}\right)^{1/2} \mu_{ij}\cdot\varepsilon_k \left\{\hat{a}_k e^{i(\mathbf{k}\cdot\mathbf{r}-\omega_k t)} - \hat{a}_k^\dagger e^{-i(\mathbf{k}\cdot\mathbf{r}-\omega_k t)}\right\} \left(\left| 1 \right\rangle \left\langle 2 \right| + \left| 2 \right\rangle \left\langle 1 \right|\right)$$

Writing out the four terms explicitly, we have

$$\hat{H}_I = i\left(\frac{\hbar\omega_k}{2\varepsilon_0 V}\right)^{1/2} \mu_{ij}\cdot\varepsilon_k \left[\begin{array}{c} \left(\hat{a}_k e^{i(\mathbf{k}\cdot\mathbf{r}-\omega_k t)} \left| 1 \right\rangle \left\langle 2 \right| + \hat{a}_k e^{i(\mathbf{k}\cdot\mathbf{r}-\omega_k t)} \left| 2 \right\rangle \left\langle 1 \right|\right) \\ -\left(\hat{a}_k^\dagger e^{-i(\mathbf{k}\cdot\mathbf{r}-\omega_k t)} \left| 1 \right\rangle \left\langle 2 \right| + \hat{a}_k^\dagger e^{-i(\mathbf{k}\cdot\mathbf{r}-\omega_k t)}\right) \left| 2 \right\rangle \left\langle 1 \right| \end{array}\right]$$

and operating each term on our product atom–field states (Eqs. 5.30, 5.31) we see that

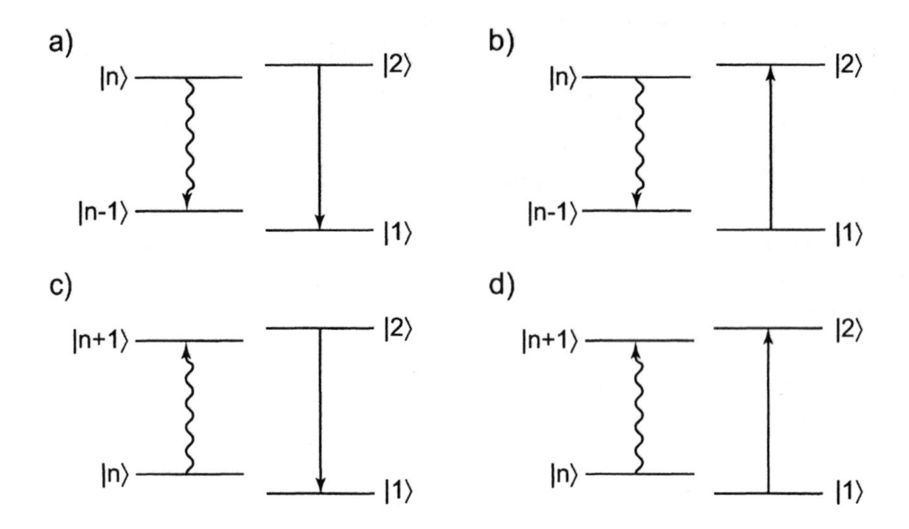

Figure 5.2: Four terms of the atom–field interaction. Terms (b) and (c) conserve energy while (a) and (d) do not.

- $|2;n\rangle \longrightarrow |1;n-1\rangle$ The atom deexcites with absorption of a photon.
- $|1;n\rangle \longrightarrow |2;n-1\rangle$ The atom is excited with the absorption of a photon.
- $|2;n\rangle \longrightarrow |1;n+1\rangle$ The atom deexcites with emission of a photon.
- $|1;n\rangle \longrightarrow |2;n+1\rangle$ The atom is excited with emission of a photon.

Obviously only the second and third terms respect energy conservation and can serve as the initial and final states of a real physical process. The first and fourth terms can be used to couple intermediate states in higher-order processes such as multiphoton absorption or Raman scattering processes. Figure 5.2 diagrams the four terms. Focusing on the second and third terms, we can simplify the notation by identifying $|2\rangle\langle 1|$ with σ^+ and $|1\rangle\langle 2|$ with σ^- from Eqs. 4.88 :

$$\hat{H}_I = i\left(\frac{\hbar\omega_k}{2\varepsilon_0 V}\right)^{1/2} \mu_{12}\cdot\varepsilon_k \left[\hat{a}_k e^{i(\mathbf{k}\cdot\mathbf{r}-\omega_k t)}\sigma^+ - \hat{a}_k^\dagger e^{-i(\mathbf{k}\cdot\mathbf{r}-\omega_k t)}\sigma^-\right]$$

Evaluating the matrix elements $\hat{H}_I$ for the general atom–field state

$$\psi = N_1|1;n-1\rangle e^{-i\omega_1 t} + N_2|1;n\rangle e^{-i\omega_1 t} + N_3|2;n-1\rangle e^{-i\omega_2 t} + N_4|2;n\rangle e^{-i\omega_2 t}$$

we see that

$$\langle\psi|\hat{H}_I|\psi\rangle = i\left(\frac{\hbar\omega_k}{2\varepsilon_0 V}\right)^{1/2} \mu_{12}\cdot\varepsilon_k \left[\begin{array}{c} N_1^* N_4 e^{i[(\mathbf{k}\cdot\mathbf{r}-(\omega_k+\omega_0)t]} + N_3^* N_2 e^{i[(\mathbf{k}\cdot\mathbf{r}-(\omega_k-\omega_0)t]} \\ -N_2^* N_3 e^{-i[(\mathbf{k}\cdot\mathbf{r}-(\omega_k-\omega_0)t]} - N_4^* N_1 e^{-i[(\mathbf{k}\cdot\mathbf{r}-(\omega_k+\omega_0)t]} \end{array}\right]$$

We see that neglecting the "unphysical" first and fourth terms is equivalent to making the rotating - wave approximation (RWA) and that the coupling between the two basis states $|1;n\rangle$ and $|2;n-1\rangle$ is really all we have to consider. The problem reduces to diagonalizing a nearly degenerate ($\Delta\omega << \omega_0$) two-level Hamiltonian operator in which the amplitude of the off-diagonal elements is given by

$$\frac{\hbar\Omega_0}{2} = \left(\frac{\hbar\omega_k}{2\varepsilon_0 V}\right)^{1/2} \mu_{12} \cdot \varepsilon_k$$

and as usual

$$\Omega = \left[(\Delta\omega)^2 + \Omega_0^2\right]^{1/2}$$

The eigenenergies of this two-level $\hat{H} = \hat{H}_A + \hat{H}_F + \hat{H}_I$ diagonalized Hamiltonian are

$$E_{\pm} = \frac{\hbar}{2}\left[\omega_{1n} + \omega_{2n-1}\right] \pm \frac{\hbar}{2}\Omega$$

where $\hbar\omega_{1n}, \hbar\omega_{2n-1}$ are the product basis state energies $\hbar\omega_1 + \hbar\omega_k n$ and $\hbar\omega_2 + \hbar\omega_k (n-1)$

5.5.2 Dressed states

The atom–field products states provide a natural set of basis states for the Hamiltonian of Eq. 5.28. The states resulting from the diagonalization of the Hamiltonian in this basis are called "dressed states". As indicated in Fig. 5.1, the closely lying doublets of the double-ladder basis 'repel' under the influence of the $\hat{H}_I$ coupling term in Eq. 5.28. The mixing coefficients reduce to the familiar two-level problem. From Fig 5.1, we obtain

$$\begin{aligned} |1, N\rangle &= \cos\theta\,|1;n\rangle + \sin\theta\,|2;n-1\rangle \\ |2, N\rangle &= \cos\theta\,|2;n-1\rangle - \sin\theta\,|1,;n\rangle \end{aligned}$$

with

$$\tan 2\theta = \frac{\Omega_0}{\Delta\omega}$$

where $\Delta\omega$ and Ω_0 have their usual meanings and the separation between members of the same dressed state manifold is

$$\hbar\Omega = \hbar\left[(\Delta\omega)^2 + \Omega_0^2\right]^{1/2}$$

5.5.3 Some applications of dressed states

Dipole gradient potential

We have seen in Section 4.4.2 how the real (dispersive) term of the susceptibility χ' interacting with the spatial gradient of the electric field amplitude E_0 can give rise to a net, period-averaged force on the atom (Eq. 4.43). The frequency

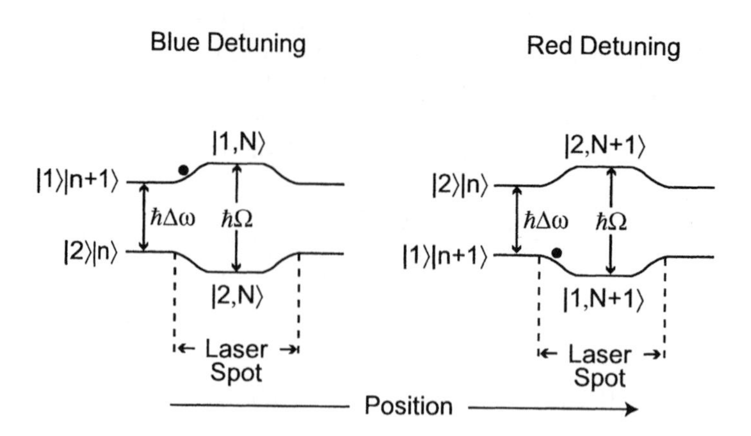

Figure 5.3: Left diagram shows product and dressed states for blue detuning. Note that population is in the upper level and the atom is subject to a low-field seeking (repulsive) force when entering the laser spot. Right diagram is similar but for red detuning. Population is in the lower level and the atom is subject to a high-field seeking (attractive) force when entering the laser spot

dependence of χ', changing sign at zero detuning, means that the resulting conservative force attracts the atom to the space of high field amplitude when the frequency is tuned below ω_0 and repels it to low field when tuned above. Integration over the relevant space coordinates results in an effective optical potential well or barrier for the atom. The qualitative behavior of the dipole gradient potential and its effect on atom motion is very easy to visualize in the dressed-states picture (see Section 6.2 and especially Eq. 6.8 for a more quantitative description). Figure 5.3 shows what happens as an atom enters a well-defined optical field space—the zone of a focused laser spot, for example. Outside the zone the atom–dipole coupling $\hbar\Omega$ is negligible and the "dressed states" are just the atom–field product states. As the atom enters the field, Ω becomes nonzero, the atom–field basis states combine to produce the dressed–state manifold, and the product–state energy levels "repel" and evolve into the dressed-state levels. Assuming that the laser is sufficiently detuned to keep the absorption rate negligible, the population remains in the ground state. We can see at a glance that blue (red) detuning leads to a repulsive (attractive) potential for the atom populated in the ground state. Furthermore, since $\hbar\Omega$ is directly proportional to the square root of the laser intensity, it is obvious that increasing this intensity (optical power per unit area) leads to a stronger force on the atom ($|F| \simeq \nabla_R \Omega$).

Problem 5.1 *Consider an external monomode (TEM_{00}) laser focused on a cloud of cold Na atoms at a temperature of 450 μK. For a detuning of 600 MHz and a focused spot diameter of 10 μm, calculate the laser intensity (W/cm^2) re-*

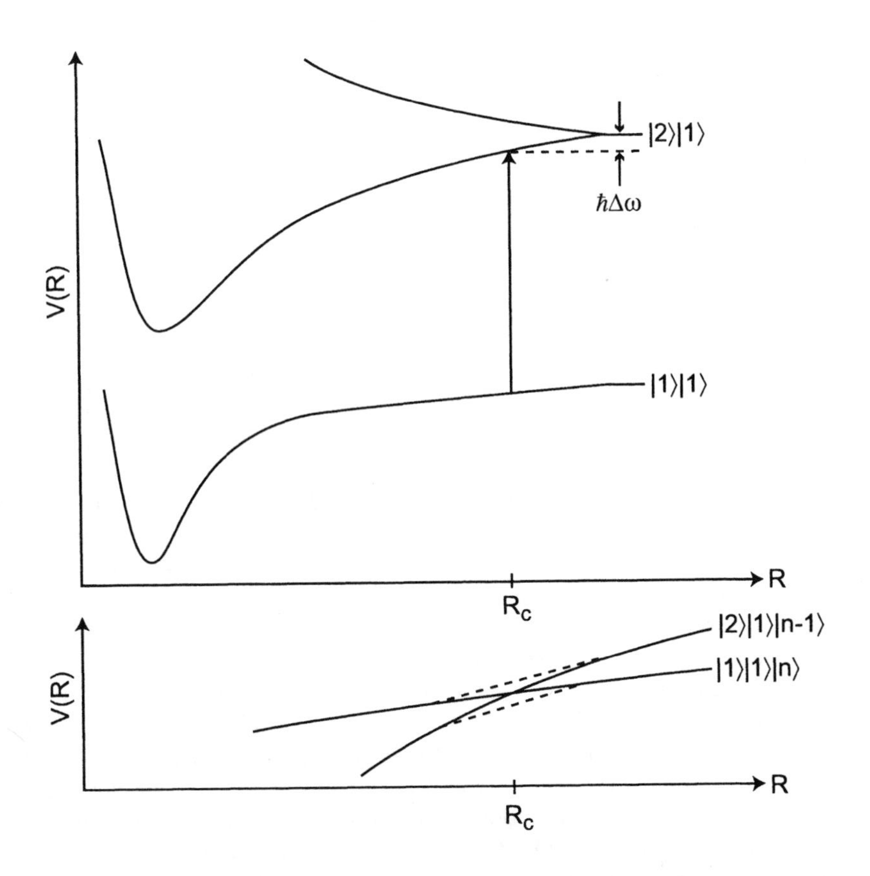

Figure 5.4: Top panel: molecular states resonantly coupled by laser field at Condon point R_C. Bottom panel: same molecule state coupling represented as an avoided crossing in the dressed - state basis.

quired to produce a potential well sufficient to contain the atoms. The transition moment [atomic units (a.u.)] of Na is 2.55.

Ultracold collisions

Ultracold collisions provide an interesting example of how light can control the outcome of inelastic or reactive collisions. Here we discuss a specific example, photoassociation, that illustrates the utility of the dressed-state point of view. The top panel of Fig. 5.4 shows the (undressed) schematic potential curves relevant to the discussion. Two ground-state atoms form a relatively flat ground molecular state characterized by the electrostatic

$$V_1(r) = -\frac{C_6}{R^6}$$

dispersion or van der Waals potential at long range. Two other molecular states arise from the interaction of an excited atom with ground-state atom. The leading interaction term is the resonant dipole–dipole potential term

$$V_{2,3}(R) = \mp \frac{C_3}{R^3}$$

which gives rise to an attractive and a repulsive potential. The inverse R^{-3} dependence of the resonant dipole interaction means that the associated potentials modify significantly the asymptotic level even at internuclear distances where the ground-state van der Waals potential is still relatively flat. In simplest terms, photoassociation involves the approach of two slow identical atoms in the ground state. An optical single-mode field, detuned to the red of the atomic resonance, is applied to the space of the colliding particles. When the two atoms approach to an internuclear distance R_C such that the applied field energy $\hbar\omega_C$ just matches the potential difference $V_2(R_C) - V_1(R_C)$, the probability to transfer population from the ground molecular state to the excited molecular is maximal. This "molecular resonance" point is sometimes called the *Condon point*. The conventional approach to calculating this probability parallels the procedure worked out in Section 2.2 for the two-level atom. First we would solve the time-independent molecular Schrödinger equation to obtain the molecular wavefunctions, then write down the coupled differential equations relating the time dependence of the expansion coefficients of the relevant molecular wavefunctions, solve for the coefficients, and take the square of their absolute value. Finally the transition probability would have to integrated over a zone ΔR to the right and left of the Condon point where the transition probability would be nonnegligible. The dressed-state picture allows this rather laborious program to be reduced to essentially a two-level curve-crossing problem. The bottom panel of Fig. 5.4 illustrates photoassociation in the dressed-states picture. The basis states are now product molecule–field states, and we approximate the molecular states themselves as products of atomic states. This approximation is justified by the long-range, weak perturbative influence of the van der Waals and resonant dipole interactions. Labeling the atom ground and excited states $|1\rangle$ and $|2\rangle$, respectively, we have

$$|1;n\rangle = |1\rangle\,|1\rangle\,|n\rangle$$

and for the field–molecule excited state

$$|2;n-1\rangle = |2\rangle\,|1\rangle\,|n-1\rangle$$

The two molecular curves intersect at the Condon point and couple optically with the applied field. This optical coupling produces an "avoided" crossing around R_C and mixing of the molecule–field basis states. The well-known and celebrated Landau–Zener (LZ) formula expresses the probability of crossing from one adiabatic molecular state to the other as a function of the strength of the interaction, the relative velocity of the colliding partners, and the relative slopes of the two curves. The LZ probability is given by

$$P_{\mathrm{LZ}} = \exp\left[\frac{2\pi\left(\langle 1;n|\,\frac{\Omega_0}{2}\,|2;n-1\rangle\right)}{v\left(\frac{d}{dR}\Delta V_{12}(R_C)\right)}\right]$$

where v is the relative radial velocity of the approaching particles and $\frac{d}{dR}\Delta V_{12}(R_C)$ is the difference in the slopes of the two noninteracting potentials at the Condon point. Dipole–field interaction operator Ω_0 should be taken with the *molecular* transition dipole. A reasonable approximation is to take the molecular transition moment as twice the atomic moment and average over all space. The result is

$$P_{\mathrm{LZmol}} = \exp\left[\frac{\frac{2\pi}{\sqrt{3}}\left(\langle 1;n|\,\Omega_{0\mathrm{at}}\,|2;n-1\rangle\right)}{v\left(\left|\frac{d}{dR}\Delta V_{12}(R_C)\right|\right)}\right]$$

where $\Omega_{0\mathrm{at}}$ denotes the atomic dipole–field interaction operator. For the case of an essentially flat V_1 potential crossing $V_2(R) = -\frac{C_3}{R^3}$ the absolute value of the derivative of the slope difference is

$$\left|\frac{d}{dR}\Delta V_{12}(R_C)\right| = \frac{3C_3}{R_C^3}$$

Problem 5.2 *Consider an external laser focused on a cloud of cold, confined Na atoms at a temperature of 450 μK. For a detuning of 600 MHz, calculate the laser intensity (W/cm²) required to produce a photoassociation probability of 25%. The transition moment (a.u.) of Na is 2.55.*

5.6 Further Reading

The treatment of field quantization and second quantization presented here is quite conventional and can be found in many books. Here are a few examples drawn from the standard references.

- R. Louden, *The Quantum Theory of Light*, 2nd edition, chapter 4, Clarendon Press, Oxford, 1983.

- M. Weissbluth, *Photon-Atom Interactions*, chapter IV, Academic Press, Boston, 1989.

- M. O. Scully, and M. Zubairy, *Quantum Optics*, chapter I, Cambridge Press, Cambridge, UK, 1997.

An excellent presentation of dressed states with a quantized field can be found in

- C. Cohen-Tannoudji, J. Dupont-Roc, G. Grynberg, *Atom-Photon Interactions,*, chapter VI, Wiley-Interscience, New York, 1992.

Semiclassical dressed states were first used by Autler and Townes to describe line doubling in molecular rotational absorption spectra under the influence of intense radio frequency excitation.

- S. H. Autler and C. H. Townes, *Phys. Rev.* **100**, 703 (1955).

In Appendix 5.A we have followed the treatment of semiclassical dressed states by Boyd in

- R. W. Boyd, *Nonlinear Optics*, chapter 5, Academic Press, Boston, 1992.

Appendix to Chapter 5

5.A Semiclassical Dressed States

Semiclassical dressed states exhibit the curious property of being stationary-state solutions of the semiclassical Schrödinger equation but are *not* energy eigenstates of the semiclassical Hamiltonian, Eq. 2.2. Because the semiclassical Hamiltonian is explicitly time-dependent, its eigenvalues must also be. This situation contrasts with the quantized field treatment where the all terms in the Hamiltonian, Eq. 5.28, are time independent, and the stationary states of the system are also eigenstates.

We return to Eq. 2.3

$$\Psi(\mathbf{r}, t) = C_1(t)\,\psi_1 e^{-i\omega_1 t} + C_2(t)\,\psi_2 e^{-i\omega_2 t} \tag{5.34}$$

and Eqs. 2.10 and 2.11 describing the time evolution of the states of our two-level atom, coupled by a classical dipole radiation field. Invoking the rotating-wave approximation, we write these two equations as

$$\frac{\Omega_0}{2} e^{i(\omega-\omega_0)t} C_2 = i\frac{dC_1}{dt} \tag{5.35}$$

and

$$\frac{\Omega_0^*}{2} e^{-i(\omega-\omega_0)t} C_1 = i\frac{dC_2}{dt} \tag{5.36}$$

We will now solve these coupled equations by first writing a trial solution for C_1 as

$$C_1 = Ke^{-i\lambda t} \tag{5.37}$$

From the second of the coupled equations, and with $\Delta\omega = \omega - \omega_0$, we have

$$\frac{dC_2}{dt} = -iKe^{-i\lambda t}\frac{\Omega_0^*}{2}e^{-i\Delta\omega t} \tag{5.38}$$

Taking the derivative of the trial solution for C_1, we obtain

$$\frac{dC_1}{dt} = -i\lambda K e^{-i\lambda t}$$

Now substituting the time derivative of the trial solution into Eq. 5.35, we can express C_2 as

$$C_2 = \frac{2\lambda K}{\Omega_0} e^{-i(\Delta\omega+\lambda)t} \tag{5.39}$$

and again taking the time derivative yields

$$\frac{dC_2}{dt} = -i(\Delta\omega + \lambda)\frac{2\lambda K}{\Omega_0} e^{-(\Delta\omega+\lambda)t}$$

Setting the right members of Eqs. 5.38 and 5.39 equal, results in a quadratic equation in the trial function parameter λ

$$\lambda^2 + (\Delta\omega)\lambda - \frac{|\Omega_0|^2}{4} = 0$$

with roots

$$\lambda = -\frac{\Delta\omega}{2} \pm \frac{1}{2}\left[(\Delta\omega)^2 + |\Omega_0|^2\right]^{1/2}$$

We define Ω in the relation

$$\Omega^2 = (\Delta\omega)^2 + |\Omega_0|^2$$

and identify Ω and Ω_0 with the *Rabi frequency* and *on-resonance Rabi frequency*, respectively, expressed earlier, in Eqs. 4.8 and 4.10. We then express the two roots λ succinctly in terms of the detuning and the Rabi frequency as

$$\lambda_{\pm} = -\frac{\Delta\omega}{2} \pm \frac{1}{2}\Omega \tag{5.40}$$

Now we substitute Eq. 5.40 into Eq. 5.37, separate the constant K into two parts, K_+ , K_- and associate each part with corresponding root, λ_+ , λ_-. The result is a general expression for the time evolution of $C_1(t)$:

$$C_1(t) = e^{i(\Delta\omega/2)t}\left[K_+ e^{-i\frac{\Omega t}{2}} + K_- e^{i\frac{\Omega t}{2}}\right] \tag{5.41}$$

Again taking the time derivative of $C_1(t)$ yields

$$\frac{dC_1}{dt} = e^{i(\Delta\omega/2)t} i\left\{K_+\left[\frac{\Delta\omega}{2} - \frac{\Omega}{2}\right]e^{-i\frac{\Omega t}{2}} + K_-\left[\frac{\Delta\omega}{2} + \frac{\Omega}{2}\right]e^{i\frac{\Omega t}{2}}\right\}$$

which, when substituted into Eq. 5.35, results in an expression for the time evolution of $C_2(t)$

$$C_2(t) = -e^{-i(\Delta\omega/2)t}\left[K_+\left(\frac{\Delta\omega - \Omega}{\Omega_0}\right)e^{-i\frac{\Omega t}{2}} + K_-\left(\frac{\Delta\omega + \Omega}{\Omega_0}\right)e^{i\frac{\Omega t}{2}}\right] \tag{5.42}$$

Now from Eq. 5.41 we can form the unnormalized probability amplitude of finding the system in the lower state

$$|C_1|^2 = K_+^2 + K_-^2 + 2K_+K_- \cos(\Omega t)$$

and if we choose some convenient time, say, $t = 0$, then we see that

$$|C_1|^2 = K_+^2 + K_-^2 + 2K_+K_- = 1$$

Clearly K_+ and K_- are not independent and must satisfy

$$K_+ + K_- = 1$$

We are free to choose the value of one of these constants. Inspection of Eq. 5.42 shows that if we take $K_+ = 1$ and $K_- = 0$, $C_2(t)$ will depend on time only in a phase factor, and consequently the probability of finding the system in the upper state $|C_2|^2$ will be time-independent:

$$|C_2|^2 = \left(\frac{\Delta\omega - \Omega}{\Omega_0}\right)^2$$

For the lower state

$$|C_1|^2 = 1$$

We write the system wavefunction, Eq. 5.34, as

$$\Psi_+ = N\left[C_1\psi_1(r)e^{-\omega_1 t} + C_2\psi_2(r)e^{-\omega_2 t}\right]$$

which, on substitution of C_2, becomes

$$\Psi_+ = N\left[\psi_1(r)e^{-i\left[\omega_1 - \frac{1}{2}(\Delta\omega+\Omega)\right]t} + \left(\frac{\Delta\omega - \Omega}{\Omega_0}\right)\psi_2(r)e^{-i\left[\omega_2 + \frac{1}{2}(\Delta\omega-\Omega)\right]t}\right]$$

with N the normalization constant. We could just as well have chosen $K_+ = 0$ and $K_- = 1$, in which case we would have

$$\Psi_- = N\left[\psi_1(r)e^{-i\left[\omega_1 - \frac{1}{2}(\Delta\omega-\Omega)\right]t} + \left(\frac{\Delta\omega + \Omega}{\Omega_0}\right)\psi_2(r)e^{-i\left[\omega_2 + \frac{1}{2}(\Delta\omega+\Omega)\right]t}\right]$$

The two possible system states can be compactly written as

$$\Psi_\pm = N\left[\psi_1(r)e^{-i\left[\omega_1 - \frac{1}{2}(\Delta\omega\pm\Omega)\right]t} + \left(\frac{\Delta\omega \mp \Omega}{\Omega_0}\right)\psi_2(r)e^{-i\left[\omega_2 + \frac{1}{2}(\Delta\omega\mp\Omega)\right]t}\right]$$

The task now is to determine the normalization constant N from

$$\int_0^\infty \Psi_\pm^* \Psi_\pm dr = 1$$

Taking into account the orthonormality of ψ_1, ψ_2 when carrying out the integrations on $\Psi_\pm$, we find

$$N_\pm = \frac{\Omega_0}{\left[\Omega_0^2 + (\Delta\omega \mp \Omega^2)\right]^{1/2}}$$

or alternatively

$$N_\pm = \frac{\Omega_0}{\Omega}\left[\frac{\Omega}{2\,(\Omega \mp \Delta\omega)}\right]^{1/2}$$

so finally the normalized system wavefunction becomes

$$\Psi_\pm = \frac{\Omega_0}{\Omega}\left[\frac{\Omega}{2\,(\Omega \mp \Delta\omega)}\right]^{1/2}\psi_1(r)e^{-i\left[\omega_1 - \frac{1}{2}(\Delta\omega\pm\Omega)\right]t} \pm \left[\frac{\Omega \mp \Delta\omega}{2\Omega}\right]^{1/2}\psi_2(r)e^{-i\left[\omega_2 + \frac{1}{2}(\Delta\omega\mp\right.}$$

We have in effect a linear combination of two new "dressed states" whose mixing coefficients are time-independent and given by

$$\widetilde{C}_1 = \frac{\Omega_0}{\Omega}\left[\frac{\Omega}{2\,(\Omega \mp \Delta\omega)}\right]^{1/2}$$

$$\widetilde{C}_2 = \pm\left[\frac{\Omega \mp \Delta\omega}{2\Omega}\right]^{1/2}$$

and whose energies are given by

$$\hbar\widetilde{\omega}_1 = \hbar\left[\omega_1 - \frac{1}{2}\left(\Delta\omega \pm \Omega\right)\right]$$
$$\hbar\widetilde{\omega}_2 = \hbar\left[\omega_2 + \frac{1}{2}\left(\Delta\omega \mp \Omega\right)\right]$$

The probability of finding the system in the upper or lower dressed states is given by

$$\left|\widetilde{C}_1\right|^2 = \left(\frac{\Omega_0}{\Omega}\right)^2\left[\frac{\Omega}{2\left(\Omega \mp \Delta\omega\right)}\right]$$
$$\left|\widetilde{C}_2\right|^2 = \left[\frac{\Omega \mp \Delta\omega}{2\Omega}\right]$$

Part II

Light–Matter Interaction: Applications

Chapter 6

Forces from Atom–Light Interaction

6.1 Introduction

A light beam carries momentum, and the scattering of light by an object produces a force on that object. This property of light was first demonstrated through the observation of a very small transverse deflection (3×10^{-5} rad) in a sodium atomic beam exposed to light from a resonance lamp. With the invention of the laser, it became easier to observe effects of this kind because the strength of the force is greatly enhanced by the use of intense and highly directional light fields. Although these results kindled interest in using light forces to control the motion of neutral atoms, the basic groundwork for the understanding of light forces acting on atoms was not laid out before the end of the 1970s. Unambiguous experimental demonstration of atom cooling and trapping was not accomplished before the mid-1980s. In this chapter we discuss some fundamental aspects of light forces and schemes employed to cool and trap neutral atoms.

The light force exerted on an atom can be of two types: a dissipative, *spontaneous force* and a conservative, *dipole force*. The spontaneous force arises from the impulse experienced by an atom when it absorbs or emits a quantum of photon momentum. As we saw in Section 2.2.6, when an atom scatters light, the resonant scattering cross section can be written as

$$\sigma_{0a} = \frac{g_1}{g_2}\frac{\pi\lambda_0^2}{2}$$

where λ_0 is the on-resonant wavelength. In the optical region of the electromagnetic spectrum the wavelengths of light are on the order of several hundreds of nanometers, so resonant scattering cross sections become quite large, $\sim 10^{-9}$ cm^2. Each photon absorbed transfers a quantum of momentum $\hbar\mathbf{k}$ to the atom in the direction of propagation. The spontaneous emission following the absorp-

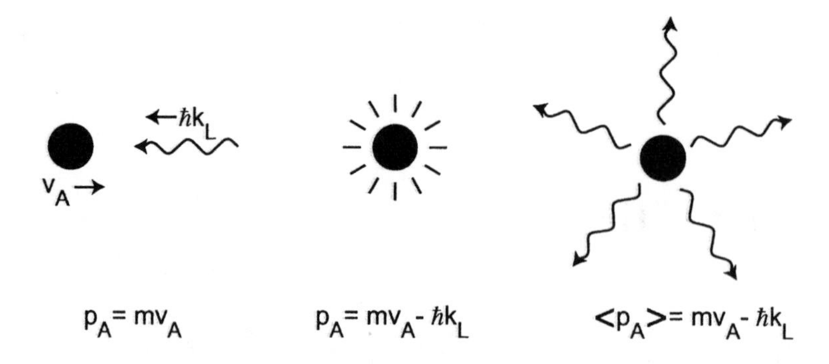

Figure 6.1: Left: atom moves to the right with mass m, velocity v_A and absorbs a photon propagating to the left with momentum $\hbar k_L$. Center: excited atom experiences a change in momentum $p_A = mv_A - \hbar k_L$. Right: photon isotropic reemission results in an average momentum change of atom, after multiple absorptions and emission, of $\langle p_A \rangle = mv_A - \hbar k_L$

tion occurs in random directions; and, over many absorption–emission cycles, it averages to zero. As a result, the *net* spontaneous force acts on the atom in the direction of the light propagation, as shown schematically in Fig. 6.1. The saturated rate of photon scattering by spontaneous emission (the reciprocal of the excited-state lifetime) fixes the upper limit to the force magnitude. This force is sometimes called *radiation pressure*.

The dipole force can be readily understood by considering the light as a classical wave. It is simply the time-averaged force arising from the interaction of the transition dipole, induced by the oscillating electric field of the light, with the gradient of the electric field amplitude. Focusing the light beam controls the magnitude of this gradient, and detuning the optical frequency below or above the atomic transition controls the sign of the force acting on the atom. Tuning the light below resonance attracts the atom to the center of the light beam, while tuning above resonance repels it. The dipole force is a stimulated process in which no net exchange of energy between the field and the atom takes place. Photons are absorbed from one mode and reappear by stimulated emission in another. Momentum conservation requires that the change of photon propagation direction from initial to final mode imparts a net recoil to the atom. Unlike the spontaneous force, there is in principle no upper limit to the magnitude of the dipole force since it is a function only of the field gradient and detuning.

6.2 The Dipole Gradient Force and the Radiation Pressure Force

We can bring these qualitative remarks into focus by considering the amplitude, phase, and frequency of a classical field interacting with an atomic transition dipole in a two-level atom. What follows immediately is sometimes called the *Doppler cooling model.* It turns out that atoms with hyperfine structure in the ground state can be cooled below the Doppler limit predicted by this model; and, to explain this unexpected sub-Doppler cooling, models involving interaction between a slowly moving atom and the polarization gradient of a standing wave have been invoked. We will sketch briefly in Section 6.3 the physics of these polarization gradient cooling mechanisms.

Remembering that the susceptibility is a density of transition dipoles, we can use Eqs. 4.43, 4.48, and 4.59, to write the basic expressions for the dipole gradient force $\mathbf{F_T}$ and the radiation pressure force $\mathbf{F_C}$ *per atom* as

$$\mathbf{F_T} = \frac{1}{4}\epsilon_0 \boldsymbol{\nabla} E_0^2 \left[\frac{\mu_{12}^2}{3\epsilon_0 \hbar}\left(-\frac{\Delta\omega}{(\Delta\omega)^2 + (\Gamma/2)^2 + \frac{1}{2}\Omega_0^2}\right)\right] \tag{6.1}$$

and

$$\mathbf{F_C} = \frac{1}{2}\epsilon_0 E_0^2 k\hat{\mathbf{k}} \left[\frac{\mu_{12}^2}{3\epsilon_0 \hbar}\left(\frac{\Gamma/2}{(\Delta\omega^2) + (\Gamma/2)^2 + \frac{1}{2}\Omega_0^2}\right)\right] \tag{6.2}$$

We use the notation $\mathbf{F_T}$ and $\mathbf{F_C}$ to indicate that the dipole gradient force (and associated potential) is often used to trap atoms, and the radiation pressure force is often used to cool them. Note that in Eqs. 6.1 and 6.2 we have used the orientation-averaged square of the transition moment matrix element, $\mu_{12}^2/3$. With the definition for the on-resonance Rabi frequency (Eq. 2.9)

$$\Omega_0 \equiv \frac{\mu_{12}E_0}{\hbar}$$

we can rewrite Eqs. 6.1 and 6.2 as

$$\mathbf{F_T} = -\frac{1}{6}\hbar\Omega_0 \boldsymbol{\nabla}\Omega_0 \left[\frac{\Delta\omega}{(\Delta\omega)^2 + (\Gamma/2)^2 + \frac{1}{2}\Omega_0^2}\right] \tag{6.3}$$

and

$$\mathbf{F_C} = \frac{1}{6}\hbar k \Gamma \left[\frac{\frac{1}{2}\Omega_0^2}{\Delta\omega^2 + (\Gamma/2)^2 + \frac{1}{2}\Omega_0^2}\right]\hat{\mathbf{k}} \tag{6.4}$$

The saturation parameter, first introduced in Eq. 4.62

$$S = \frac{\frac{1}{2}\Omega_0^2}{(\Delta\omega)^2 + (\Gamma/2)^2}$$

allows the dipole-gradient force and the radiation pressure force to be written

$$\mathbf{F_T} = -\frac{1}{6}\hbar\Delta\omega\boldsymbol{\nabla}S \cdot \frac{1}{1+S} \tag{6.5}$$

and

$$\mathbf{F_C} = \frac{1}{6}\hbar\mathbf{k}\Gamma \cdot \frac{S}{1+S} \tag{6.6}$$

Equation 6.6 shows that the radiation pressure force "saturates" as S increases; and is therefore limited by the spontaneous emission rate. The saturation parameter essentially defines an index for the relative importance of the terms that appear in the denominator of the lineshape function for the atom–light forces. The spontaneous emission rate is an intrinsic property of the atom, proportional to the square of the atomic transition moment, while the square of the Rabi frequency is a function of the exciting laser intensity. If $S \ll 1$, the spontaneous emission rate is fast compared to any stimulated process, and the exciting light field is said to be weak. If $S \gg 1$, the Rabi oscillation is fast compared to spontaneous emission and the field is considered strong. Setting S equal to unity defines a "saturation" condition for the transition,

$$\Omega_{\mathrm{sat}} = \sqrt{2}\left(\frac{\Gamma}{2}\right) \tag{6.7}$$

and the line-shape factor indicates a saturation "power broadening" of a factor of $\sqrt{2}$.

The dipole gradient force $\mathbf{F_T}$ can be integrated to define an attractive (or repulsive) potential for the atom:

$$U_T = -\int \mathbf{F_T}d\mathbf{r} = \frac{\hbar\Delta\omega}{6}\ln\left[1 + \frac{\frac{1}{2}{\Omega_0}^2}{(\Delta\omega)^2 + (\Gamma/2)^2}\right] \tag{6.8}$$

or in terms of the saturation parameter

$$U_T = -\int \mathbf{F_T}d\mathbf{r} = \frac{\hbar\Delta\omega}{6}\ln\left[1+S\right] \tag{6.9}$$

Note that the dipole gradient force and potential (Eqs. 6.5, 6.9) do not saturate with increasing light-field intensity. Usually $\mathbf{F_T}$ and U_T are used to manipulate and trap atoms with a laser light source detuned far from resonance to avoid absorption. In this case $S \ll 1$, and the trapping potential can be written

$$U_T \simeq \frac{1}{6}\frac{\hbar\Omega_0^2}{2\hbar\Delta\omega}$$

Often the transition moment can be oriented by using circularly polarized light. In that case all the previous expressions for $\mathbf{F_T}, \mathbf{F_C}$, and U_T must be multiplied by 3. From now on we will drop the orientation averaging and just use μ_{12}^2 for the square of the transition moment.

From the previous definitions of I, Ω_0, and Ω_{sat}, (Eqs. 1.10, 2.9, 6.7), we can write

$$\frac{I}{I_{\mathrm{sat}}} = \frac{\Omega_0^2}{\Gamma^2/2} \tag{6.10}$$

and

$$\mathbf{F_C} = \frac{\hbar \mathbf{k} \Gamma}{2} \left[\frac{I/I_{\text{sat}}}{\left(\frac{2\Delta\omega}{\Gamma}\right)^2 + I/I_{\text{sat}} + 1} \right] \tag{6.11}$$

Now if we consider the atom moving in the $+z$ direction with velocity v_z and counterpropagating to the light wave detuned from resonance by $\Delta\omega_L$, the *net* detuning will be

$$\Delta\omega = \Delta\omega + kv_z \tag{6.12}$$

where the term kv_z is the Doppler shift. The force F_- acting on the atom will be in the direction opposite to its motion. In general

$$\mathbf{F}_\pm = \pm \frac{\hbar \mathbf{k} \Gamma}{2} \left[\frac{I/I_{\text{sat}}}{\left(\frac{2(\Delta\omega \mp kv_z)}{\Gamma}\right)^2 + I/I_{\text{sat}} + 1} \right] \tag{6.13}$$

Suppose we have two fields propagating in the $\pm z$ directions and we take the net force $\mathbf{F} = \mathbf{F}_+ + \mathbf{F}_-$. If kv_z is small compared to Γ and $\Delta\omega$, then we find

$$\mathbf{F} \simeq 4\hbar \mathbf{k} \frac{I}{I_{\text{sat}}} \frac{kv_z(2\Delta\omega/\Gamma)}{\left[1 + \frac{I}{I_{\text{sat}}} + (2\Delta\omega/\Gamma)^2\right]^2} \tag{6.14}$$

This expression shows that if the detuning $\Delta\omega$ is negative (i.e., red-detuned from resonance), then the cooling force will oppose the motion and be proportional to the atomic velocity. Figure 6.2 plots this dissipative restoring force as a function of v_z at a detuning $\Delta\omega = -\Gamma$ and $I/I_{\text{sat}} = 2$. The one-dimensional motion of the atom, subject to an opposing force proportional to its velocity, is described by a damped harmonic oscillator. The Doppler damping or friction coefficient is the proportionality factor,

$$\alpha_d = -4\hbar k^2 \frac{I}{I_{\text{sat}}} \frac{(2\Delta\omega/\Gamma)}{\left[1 + \frac{I}{I_{\text{sat}}} + (2\Delta\omega/\Gamma)^2\right]^2} \tag{6.15}$$

and the characteristic time to damp the kinetic energy of the atom of mass m to $1/e$ of its initial value is,

$$\tau = \frac{m}{2\alpha_d} \tag{6.16}$$

However, the atom will not cool indefinitely. At some point the Doppler cooling rate will be balanced by the heating rate coming from the momentum fluctuations of the atom absorbing and reemitting photons. Setting these two rates equal and associating the one-dimensional kinetic energy with $\frac{1}{2}k_B T$, we find

$$k_B T = \frac{\hbar\Gamma}{4} \frac{1 + (2\Delta\omega/\Gamma)^2}{2|\Delta\omega|/\Gamma} \tag{6.17}$$

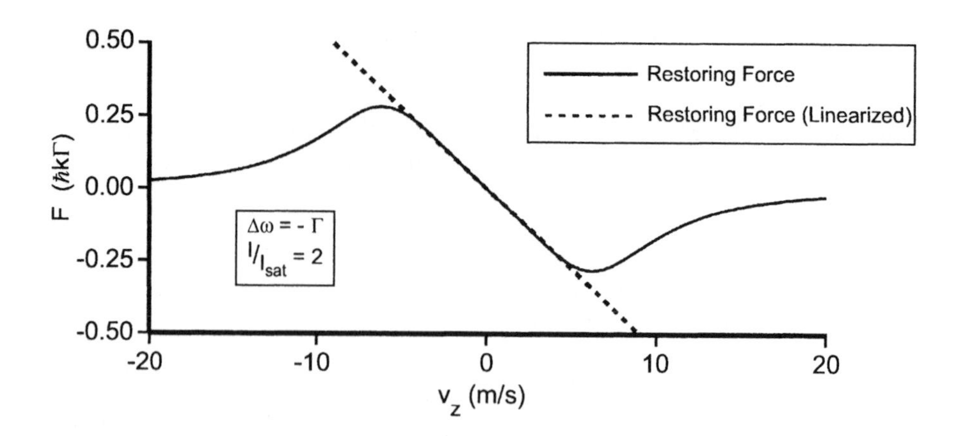

Figure 6.2: One-dimensional Doppler radiation pressure force versus atom velocity along z axis for a red detuning of one natural line width and a light intensity of $2I_{\rm sat}$. The solid line plots the exact expression for the restoring force (Eq. 6.13). The dashed line plots the approximate expression (linear in velocity dependence) of Eq. 6.14.

This expression shows that T is a function of the laser detuning, and the minimum temperature is obtained when $\Delta\omega = -\frac{\Gamma}{2}$. At the this detuning

$$k_B T = \hbar \frac{\Gamma}{2} \tag{6.18}$$

which is called the *Doppler-cooling limit.* This limit is typically, for alkali atoms, on the order of a few hundred microkelvins. For example, the Doppler cooling limit for Na is $T = 240\mu$K. In the early years of cooling and trapping, prior to 1988, the Doppler limit was thought to be a real physical barrier, but in that year several groups showed that in fact Na atoms could be cooled well below the Doppler limit. Although the physics of this sub-Doppler cooling in three dimensions is still not fully understood, the essential role played by the hyperfine structure of the ground state has been worked out in one-dimensional models, which we describe in the following section.

6.3 Sub-Doppler Cooling

Two principal mechanisms that cool atoms to temperatures below the Doppler limit rely on spatial polarization gradients of the light field through which the atoms move. These two mechanisms, however, invoke very different physics, and are distinguished by the spatial polarization dependence of the light field. A key point is that these sub-Doppler mechanisms operate only on multilevel atoms; and, in particular, it is essential to have multiple levels in the ground state.

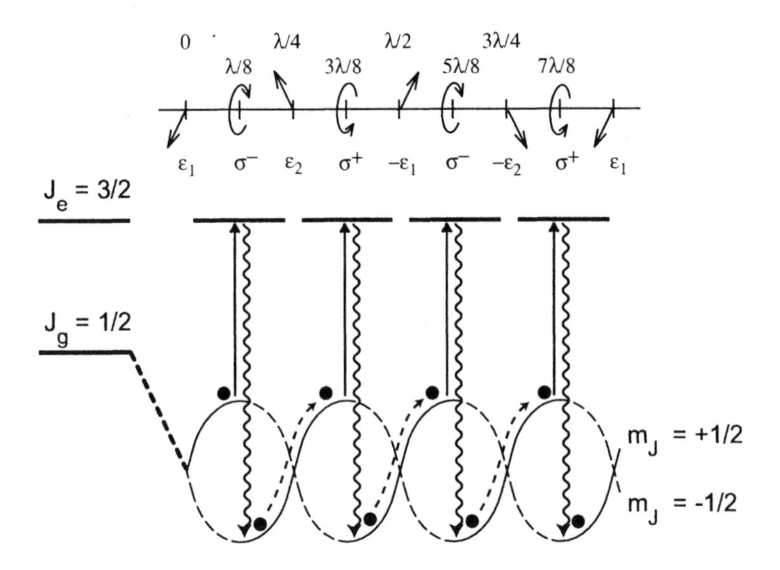

Figure 6.3: Upper line shows the change in polarization as a function of distance (in wavelength units) for the "lin-perp-lin" (linear–perpendicular–linear) standing wave configuration. Lower figure shows a simplified schematic of the Sisyphus cooling mechanism for the $J_{1/2} \longleftrightarrow J_{3/2}$ two-level atom.

Therefore, strictly speaking, the subject of sub-Doppler cooling lies outside the scope of the two-level atom. Nevertheless, because of its importance for real cooling in the alkali atoms, for example, we include it here. Two parameters, the friction coefficient and the velocity capture range, determine the significance of these cooling processes. In this section we compare expressions for these quantities in the sub-Doppler regime to those found in the conventional Doppler cooling model of one-dimensional optical molasses.

In the first case two counterpropagating light waves with orthogonal polarization form a standing wave. This arrangement is commonly called the "lin-perp-lin" configuration. Figure 6.3 shows what happens. We see from the figure that if we take as a starting point a position where the light polarization is linear (ϵ_1), it evolves from linear to circular over a distance of $\lambda/8$ (σ_-). Then over the next $\lambda/8$ interval the polarization again changes to linear but in the direction orthogonal to the first (ϵ_2). Then from $\lambda/4$ to $3\lambda/8$ the polarization again becomes circular but in the sense opposite (σ_+) to the circular polarization at $\lambda/8$, and finally after a distance of $\lambda/2$ the polarization is again linear but out of phase with respect to (ϵ_1). Over the same half-wavelength distance of the polarization period, atom–field coupling produces a periodic energy (or light) shift in the hyperfine levels of the atomic ground state. To illustrate the cooling mechanism we assume the simplest case, a $J_g = \frac{1}{2} \rightarrow J_e = \frac{3}{2}$ transition.

As shown in Fig. 6.3, the atom moving through the region of z around $\lambda/8$, where the polarization is primarily σ_-, will have its population pumped mostly into $J_g = -\frac{1}{2}$. Furthermore the Clebsch–Gordan coefficients controlling the transition dipole coupling to $J_e = \frac{3}{2}$ impose that the $J_g = -\frac{1}{2}$ level couples to σ_- light 3 times more strongly than does the $J_g = +\frac{1}{2}$ level. The difference in coupling strength leads to the light shift splitting between the two ground state shown in Fig. 6.3. As the atom continues to move to $+\ z$, the relative coupling strengths are reversed around $3\lambda/8$, where the polarization is essentially σ_+. Thus the relative energy levels of the two hyperfine ground states oscillate "out of phase" as the atom moves through the standing wave. The key idea is that the optical pumping rate, always redistributing population to the lower-lying hyperfine level, lags the light shifts experienced by the two atom ground-state components as the atom moves through the field. The result is a "Sisyphus effect" where the atom cycles through a period in which the effectively populated atomic sublevel spends most of its time climbing a potential hill, converting kinetic energy to potential energy, subsequently dissipating the accumulated potential energy, by spontaneous emission, into the empty modes of the radiation field, and simultaneously transferring population back to the lower-lying of the two ground-state levels. The lower diagram in Fig. 6.3 illustrates the optical pumping phase lag. In order for this cooling mechanism to work, the optical pumping time, controlled by the light intensity, must be less than the light-shift time, controlled essentially by the velocity of the atom. Since the atom is moving slowly, having been previously cooled by the Doppler mechanism, the light field must be weak in order to slow the optical pumping rate so that it lags behind the light-shift modulation rate. This physical picture combines the conservative optical dipole force, whose space integral gives rise to the potential hills and valleys over which the atom moves and the irreversible energy dissipation of spontaneous emission required to achieve cooling. We can make the discussion more precise and obtain simple expressions for the friction coefficient and velocity capture by establishing some definitions. As in the Doppler cooling model we define the friction coefficient α_{lpl} to be the proportionality constant between the force F and the atomic velocity v.

$$F = -\alpha_{\mathrm{lpl}} v \tag{6.19}$$

We assume that the light field is detuned to the red of the $J_g \to J_e$ atomic resonance frequency,

$$\Delta\omega_L = \omega - \omega_0 \tag{6.20}$$

and term the light shifts of the $J_g = \pm\frac{1}{2}$ levels $\Delta_\pm$, respectively. At the position $z = \lambda/8$, $\Delta_- = 3\Delta_+$ and at $z = 3\lambda/8$, $\Delta_+ = 3\Delta_-$. Since the applied field is red-detuned, all Δ terms have negative values. Now, in order for the cooling mechanism to be effective, the optical pumping time τ_p should be comparable to the time required for the atom with velocity v to travel from the bottom to the top of a potential hill, $\frac{\lambda/4}{v}$

$$\tau_p \simeq \frac{\lambda/4}{v} \tag{6.21}$$

or

$$\Gamma' \simeq kv \tag{6.22}$$

where $\Gamma' = 1/\tau_p$ and $\lambda/4 \simeq 1/k$, with $k = \frac{2\pi}{\lambda}$ the magnitude of the optical wave vector. Now the amount of energy W dissipated in one cycle of hill climbing and spontaneous emission is essentially the average energy splitting of the two light-shifted ground states, between, say, $z = \lambda/8$ and $3\lambda/8$ or $W \simeq -\hbar\Delta$. Therefore the *rate* of energy dissipation is

$$\frac{dW}{dt} = -\Gamma'\hbar\Delta \tag{6.23}$$

But in general the time-dependent energy change of a system can be always be expressed as $\frac{dW}{dt} = \mathbf{F} \cdot \mathbf{v}$, so in this one-dimensional model and taking into account Eq. 6.19, we can write

$$\frac{dW}{dt} = -\alpha_{\rm lpl} v^2 = -\Gamma'\hbar\Delta \tag{6.24}$$

so that

$$\alpha_{\rm lpl} = -\frac{k\hbar\Delta}{v} = -\frac{k^2\hbar\Delta}{\Gamma'} \tag{6.25}$$

Note that since $\Delta < 0$, α_{lpl} is a positive quantity. Note also that at far detunings ($\Delta\omega >> \Gamma$), Eq. 6.8 shows that

$$\frac{U}{\hbar} = \Delta = \frac{\Omega_0^2}{4\Delta\omega_L}$$

Problem 6.1 *Verify that in the limit of large detuning Eq. 6.8* $\rightarrow \frac{\Omega_0^2}{4\Delta\omega}$.

It is also true that for light shifts large compared to the ground-state natural line width ($\Delta >> \Gamma'$), and at detunings far to the red of resonance ($\Delta\omega_L \gtrsim 4\Gamma$)

$$\frac{\Gamma}{\Gamma'} = \frac{\Delta\omega_L^2}{\Omega_0^2}$$

so the sub-Doppler friction coefficient can also be written

$$\alpha_{\rm lpl} = -\frac{k^2\hbar\Delta\omega_L}{4\Gamma} \tag{6.26}$$

Equation 6.26 yields two remarkable predictions: first, that the sub-Doppler "lin-perp-lin" friction coefficient can be a big number compared to α_d. Note that from Eq. 6.15, with $I \lesssim I_{\rm sat}$ and $\Delta\omega_L \gg \Gamma$, we obtain

$$\alpha_d \simeq \frac{1}{2}\hbar k^2 \left(\frac{\Gamma}{\Delta\omega_L}\right)^3$$

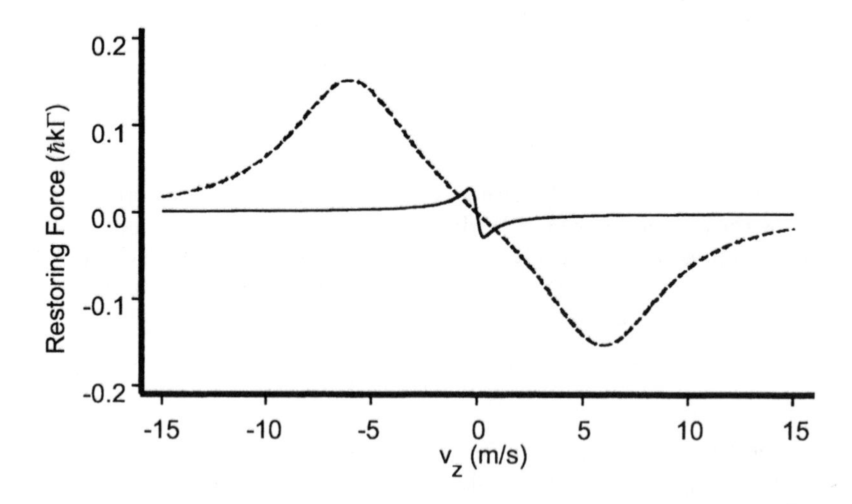

Figure 6.4: Comparison of slope, amplitude, and "capture range" of Doppler cooling and Sisyphus cooling.

and

$$\frac{\alpha_{\mathrm{lpl}}}{\alpha_d} \simeq \frac{1}{2}\left(\frac{\Delta\omega_L}{\Gamma}\right)^4$$

and second, that α_{lpl} is independent of the applied field intensity. This last result differs from the Doppler friction coefficient, which is proportional to field intensity up to saturation (*cf.* Eq. 6.15). However, even though α_{lpl} looks impressive, the range of atomic velocities over which it can operate is restricted by the condition that

$$\Gamma' \simeq kv$$

The ratio of the capture velocities for Doppler/sub-Doppler cooling is therefore only

$$v_{\mathrm{lpl}}/v_d \simeq \frac{4\Delta}{\Delta\omega_L}$$

Figure 6.4 illustrates graphically the comparison between the Doppler and the "lin-perp-lin" sub-Doppler cooling mechanism. The dramatic difference in capture range is evident from the figure. Note also that the slopes of the curves give the friction coefficients for the two regimes and that within the narrow velocity capture range of its action, the slope of the sub-Doppler mechanism is markedly steeper.

The second mechanism operates with the two counterpropagating beams circularly polarized in opposite senses. When the two counterpropagating beams have the same amplitude, the resulting polarization is always linear and orthogonal to the propagation axis, but the tip of the polarization axis traces out a helix with a pitch of λ. Figure 6.5 illustrates this case. The physics of the sub-Doppler mechanism does not rely on hill climbing and spontaneous emission,

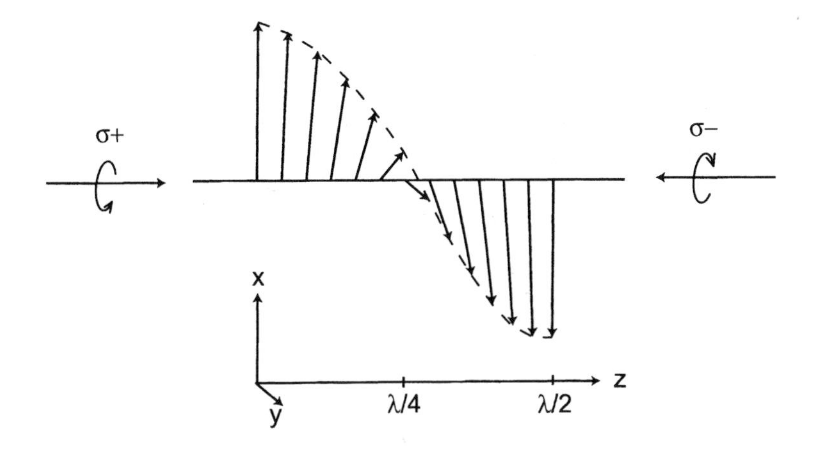

Figure 6.5: Polarization as a function of distance (in wavelength units) for the σ^+, σ^- standing-wave configuration.

but on an imbalance in the photon scattering rate from the two counterpropagating light waves as the atom moves along the z axis. This imbalance leads to a velocity-dependent restoring force acting on the atom. The essential factor leading to the differential scattering rate is the creation of *population orientation* along the z axis among the sublevels of the atom *ground state*. Those sublevels with more population scatter more photons. Now it is evident from a consideration of the energy level diagram (see Fig. 6.3) and the symmetry of the Clebsch–Gordan coefficients that $J_g = \frac{1}{2} \leftrightarrow J_e = \frac{3}{2}$ transitions coupled by linearly polarized light cannot produce a population orientation in the ground state. In fact the simplest system to exhibit this effect is $J_g = 1 \leftrightarrow J_e = 2$, and a measure of the orientation is the magnitude of the $\langle J_z \rangle$ matrix element between the $J_{g_z} = \pm 1$ sublevels. If the atom remained stationary at $z = 0$, interacting with the light polarized along y, the light shifts Δ_0, $\Delta_{\pm 1}$ of the three ground state sublevels would be

$$\Delta_{+1} = \Delta_{-1} = \frac{3}{4}\Delta_0 \tag{6.27}$$

and the steady-state populations $\frac{4}{17}$, $\frac{4}{17}$, and $\frac{9}{17}$ respectively. Evidently linearly polarized light will not produce a net steady-state orientation, $\langle J_z \rangle$. As the atom begins to move along z with velocity v, however, it sees a linear polarization precessing around its axis of propagation with an angle $\varphi = -kz = -kvt$. This precession gives rise to a new term in the Hamiltonian, $V = kvJ_z$. Furthermore, if we transform to a rotating coordinate frame, the eigenfunctions belonging to the Hamiltonian of the moving atom in this new "inertial" frame become linear combinations of the basis functions with the atom at rest. Evaluation of the steady-state orientation operator J_z in the inertial frame is now nonzero:

$$\langle J_z \rangle = \frac{40}{17}\frac{\hbar kv}{\Delta_0} = \hbar\,[\Pi_{+1} - \Pi_{-1}] \tag{6.28}$$

Notice that the orientation measure is nonzero only when the atom is moving. In Eq. 6.28 we denote the populations of the $|\pm\rangle$ sublevels as $\Pi_{\pm}$,and we interpret the nonzero matrix element as a direct measure of the population difference between the $|\pm\rangle$ levels of the ground state. Note that since Δ_0 is a negative quantity (red detuning), Eq. 6.28 tells us that the Π_- population is greater than the Π_+ population. Now if the atom traveling in the $+z$ direction is subject to two light waves, one with polarization σ_- (σ_+) propagating in the $-z\,(+z)$ direction, the preponderance of population in the $|-\rangle$ level will result in a higher scattering rate from the wave traveling in the $-z$ direction. Therefore the atom will be subject to a net force opposing its motion and proportional to its velocity. The differential scattering rate is

$$\frac{40}{17}\frac{kv}{\Delta_0}\Gamma'$$

and with an $\hbar k$ momentum quantum transferred per scattering event, the net force is

$$F = \frac{40}{17}\frac{\hbar k^2 v\Gamma'}{\Delta_0} \tag{6.29}$$

The friction coefficient α_{cp} is evidently

$$\alpha_{\mathrm{cp}} = -\frac{40}{17}\hbar k^2\frac{\Gamma'}{\Delta_0} \tag{6.30}$$

which is a positive quantity since Δ_0 is negative from red detuning. Contrasting α_{cp} with α_{lpl} we see that α_{cp} must be much smaller since the assumption has been all along that the light shifts Δ were much greater than the line widths Γ'. It turns out, however, that the heating rate from recoil fluctuations is also much smaller so that the ultimate temperatures reached from the two mechanisms are comparable.

Although the Doppler cooling mechanism also depends on a scattering imbalance from oppositely traveling light waves, the imbalance in the scattering rate comes from a difference in the scattering probability per photon due to the Doppler shift induced by the moving atom. In the sub-Doppler mechanism the scattering probabilities from the two light waves are equal but the ground-state populations are not. The state with the greater population experiences the greater rate.

6.4 The Magneto-optical Trap (MOT)

6.4.1 Basic notions

When first considering the basic idea of particle confinement by optical forces one has to confront a seemingly redoubtable obstacle—the optical Earnshaw theorem (OET). This theorem states that if a force is proportional to the light intensity, its divergence must be null because the divergence of the Poynting vector, which expresses the directional flow of intensity, must be null through a volume without sources or sinks of radiation. This null divergence rules out the possibility of an inward restoring force everywhere on a closed surface. However, the OET can be circumvented by a clever trick. When the internal degrees of freedom (i.e. energy levels) of the atom are taken into account, they can change the proportionality between the force and the Poynting vector in a position-dependent way such that the OET does not apply. Spatial confinement is then possible with spontaneous light forces produced by counterpropagating optical beams. The trap configuration that is presently the most commonly employed uses a radial magnetic field gradient produced by a quadrupole field and three pairs of circularly polarized, counterpropagating optical beams, detuned to the red of the atomic transition and intersecting at right angles at the point where the magnetic field is zero. This magneto - optical trap (MOT) exploits the position-dependent Zeeman shifts of the electronic levels when the atom moves in the radially increasing magnetic field. The use of circularly polarized light, red-detuned by about one Γ results in a spatially dependent transition probability whose net effect is to produce a restoring force that pushes the atom toward the origin.

To explain how this trapping scheme works, consider a two-level atom with a $J = 0 \rightarrow J = 1$ transition moving along the z direction. We apply a magnetic field $B(z)$ increasing linearly with distance from the origin. The Zeeman shifts of the electronic levels are position-dependent

$$\Delta\omega_B = \frac{\mu_B}{\hbar} \cdot \frac{dB}{dz} \cdot z$$

where μ_B is the Zeeman constant for the net shift of the transition frequency in the magnetic field. The Zeeman shifts are shown schematically in Fig. 6.6. We also apply counterpropagating optical fields along the $\pm z$ directions carrying oppositely circular polarization and detuned to the red of the atomic transition. It is clear from Fig. 6.6 that an atom moving along $+z$ will scatter σ^- photons at a faster rate than σ^+ photons because the Zeeman effect will shift the $\Delta M_J = -1$ transition closer to the light frequency. The expression for the radiation pressure force, which extends Eq. 6.2 to include the Doppler shift kv_z and the Zeeman shift, becomes

$$F_{1z} = -\frac{\hbar k}{2}\Gamma \frac{\frac{1}{2}\left|\Omega_0\right|^2}{\left(\Delta\omega + kv_z + \frac{\mu_B}{\hbar} \cdot \frac{dB}{dz} \cdot z\right)^2 + (\Gamma/2)^2 + \frac{1}{2}\left|\Omega_0\right|^2} \tag{6.31}$$

In a similarly way, if the atom moves along $-z$ it will scatter σ^+ photons at a faster rate from the $\Delta M_J = +1$ transition.

$$F_{2z} = +\frac{\hbar k}{2}\Gamma \frac{\frac{1}{2}\left|\Omega_0\right|^2}{\left(\Delta\omega - kv_z - \frac{\mu_B}{\hbar} \cdot \frac{dB}{dz} \cdot z\right)^2 + (\Gamma/2)^2 + \frac{1}{2}\left|\Omega_0\right|^2} \tag{6.32}$$

The atom will therefore experience a net restoring force pushing it back to the origin. If the light beams are red-detuned $\sim \Gamma$, then the Doppler shift of the atomic motion will introduce a velocity-dependent term to the restoring

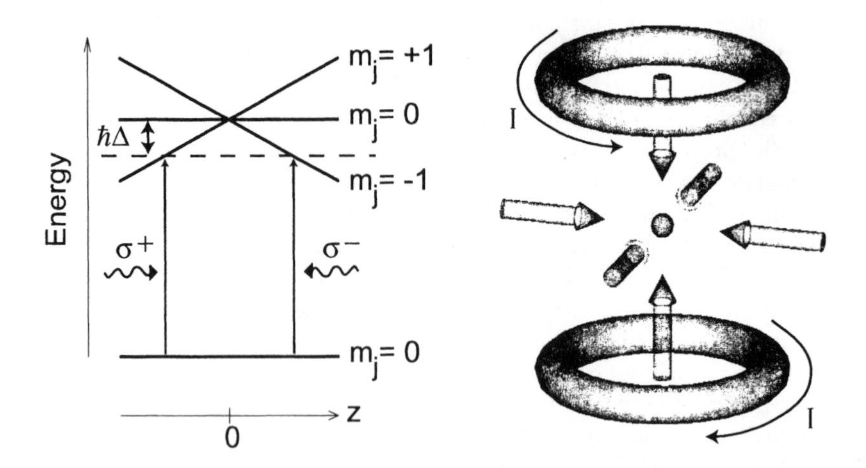

Figure 6.6: Left: diagram of the Zeeman shift of energy levels in a MOT as an atom moves to away from the trap center. Restoring force becomes localized around resonance positions as indicated. Right: schematic of a typical MOT setup showing six laser beams and antiHelmholtz configuration producing quadrupole magnetic field.

force such that, for small displacements and velocities, the total restoring force can be expressed as the sum of a term linear in velocity and a term linear in displacement[1],

$$F_{\mathrm{MOT}} = F_{1z} + F_{2z} = -\alpha\dot{z} - Kz \tag{6.33}$$

From Eq. 6.33 we can derive the equation of motion of a damped harmonic oscillator with mass m:

$$\ddot{z} + \frac{2\alpha}{m}\dot{z} + \frac{K}{m}z = 0 \tag{6.34}$$

The damping constant α and the spring constant K can be written compactly in terms of the atomic and field parameters as

$$\alpha = \hbar k\Gamma \frac{16\,|\Delta'|\,(\Omega')^2\,(k/\Gamma)}{\left[1 + 2\,(\Omega')^2\right]^2 \left[1 + \frac{4(\Delta')^2}{1+2(\Omega')^2}\right]^2} \tag{6.35}$$

and

$$K = \hbar k\Gamma \frac{16\,|\Delta'|\,(\Omega')^2\left(\frac{d\omega_0}{dz}\right)}{\left[1 + 2\,(\Omega')^2\right]^2 \left[1 + \frac{4(\Delta')^2}{1+2(\Omega')^2}\right]^2} \tag{6.36}$$

[1]Note that this development makes the tacit assumption that it is permissable to add intensities, not fields. Strictly speaking, if phase coherence is preserved between the counter-propagating beams, the fields should be added and standing waves will appear in the MOT zone. For many practical MOT setups, however, the phase-incoherent treatment is sufficient.

where Ω', Δ', and

$$\frac{\omega_0}{dz} = \frac{\left[(\mu_B/\hbar)\left(\frac{dB}{dz}\right)\right]}{\Gamma}$$

are Γ normalized analogues of the quantities defined earlier. Typical MOT operating conditions fix $\Omega' = \frac{1}{2}, \Delta' = 1$, so α and K reduce to

$$\alpha \simeq (0.132)\,\hbar k^2 \tag{6.37}$$

and

$$K \simeq \left(1.16 \times 10^{10}\right) \hbar k \cdot \frac{dB}{dz} \tag{6.38}$$

The extension of these results to three dimensions is straightforward if one takes into account that the quadrupole field gradient in the z direction is twice the gradient in the x, y directions, so that $K_z = 2K_x = 2K_y$. The velocity dependent damping term implies that kinetic energy E dissipates from the atom (or collection of atoms) as

$$\frac{E}{E_0} = e^{-\frac{2\alpha}{m}t}$$

where m is the atomic mass and E_0 the kinetic energy at the beginning of the cooling process. Therefore the dissipative force term cools the collection of atoms as well as combining with displacement term to confine them. The damping time constant

$$\tau = \frac{m}{2\alpha}$$

is typically tens of microseconds. It is important to bear in mind that a MOT is anisotropic since the restoring force along the z axis of the quadrupole field is twice the restoring force in the x–y plane. Furthermore a MOT provides a dissipative rather than a conservative trap, and it is therefore more accurate to characterize the maximum capture velocity rather than the trap "depth".

Early experiments with MOT-trapped atoms were carried out initially by slowing an atomic beam to load the trap. Later a continuous uncooled source was used for that purpose, suggesting that the trap could be loaded with the slow atoms of a room-temperature vapor. The next advance in the development of magneto-optical trapping was the introduction of the vapor-cell magneto-optical trap (VCMOT). This variation captures cold atoms directly from the low-velocity edge of the Maxwell–Boltzmann distribution always present in a cell background vapor. Without the need to load the MOT from an atomic beam, experimental apparatuses became simpler; and now many groups around the world use the VCMOT for applications ranging from precision spectroscopy to optical control of reactive collisions.

6.4.2 Densities in a MOT

The VCMOT typically captures about a million atoms in a volume less than a millimeter diameter, resulting in densities $\sim 10^{10}$ cm^{-3}. Two processes limit the density attainable in a MOT: (1) collisional trap loss and (2) repulsive forces between atoms caused by reabsorption of scattered photons from the interior of the trap. Collisional loss in turn arises from two sources: hot background atoms that knock cold atoms out of the MOT by elastic impact, and binary encounters between the cold atoms themselves. The "photon-induced repulsion" or photon trapping arises when an atom near the MOT center spontaneously emits a photon that is reabsorbed by another atom before the photon can exit the MOT volume. This absorption results in an increase of $2\hbar k$ in the relative momentum of the atomic pair and produces a repulsive force proportional to the product of the absorption cross section for the incident light beam and scattered fluorescence. When this outward repulsive force balances the confining force, further increase in the number of trapped atoms leads to larger atomic clouds, but not to higher densities.

6.4.3 Dark SPOT (spontaneous-force optical trap)

In order to overcome the "photon-induced repulsion" effect, atoms can be optically pumped to a "dark" hyperfine level of the atom ground state that does not interact with the trapping light. In a conventional MOT one usually employs an auxiliary "repumper" light beam, copropagating with the trapping beams but tuned to a neighboring transition between hyperfine levels of ground and excited states. The repumper recovers population that leaks out of the cycling transition between the two levels used to produce the MOT. As an example Fig. 6.7 shows the trapping and repumping transitions usually employed in a Na MOT. The scheme, known as a dark spontaneous-force optical trap (dark SPOT), passes the repumper through a glass plate with a small black dot shadowing the beam such that the atoms at the trap center are not coupled back to the cycling transition but spend most ($\sim 99\%$) of their time in the "dark" hyperfine level. Cooling and confinement continue to function on the periphery of the MOT but the center core experiences no outward light pressure. The dark SPOT increases density by almost two-orders of magnitude.

6.4.4 Far off-resonance trap (FORT)

Although a MOT functions as a versatile and robust "reaction cell" for studying cold collisions, light frequencies must tune close to atomic transitions, and an appreciable steady-state fraction of the atoms remain excited. Excited-state trap-loss collisions and photon-induced repulsion limit achievable densities.

A far-off-resonance trap (FORT), in contrast, uses the dipole force rather than the spontaneous force to confine atoms and can therefore operate far from resonance with negligible population of excited states. The FORT consists of a single, linearly polarized, tightly focused Gaussian mode beam tuned far to the red of resonance. The obvious advantage of large detunings is the suppression of photon absorption. Note from Eq. 6.2 that the spontaneous force, involving absorption and reemission, falls off as the square of the detuning, while Eq. 6.8 shows that the potential derived from dipole force falls off only as the detuning

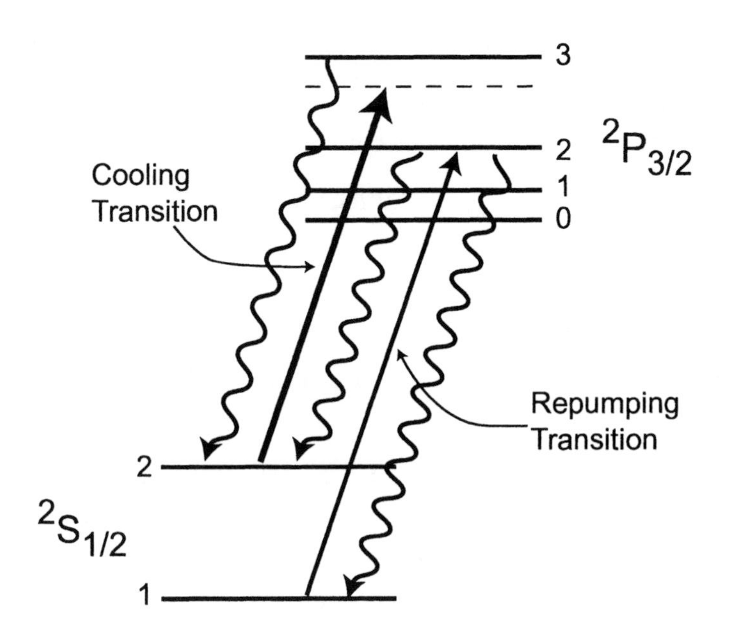

Figure 6.7: Hyperfine structure in sodium atom showing the usual cooling, pumping, and repumping transitions.

itself. At large detunings and high field gradients (tight focus) Eq. 6.8 becomes

$$U \simeq \frac{\hbar\,|\Omega_0|^2}{4\Delta\omega} \tag{6.39}$$

which shows that the potential becomes directly proportional to light intensity and inversely proportional to detuning. Therefore, at far detuning but high intensity the depth of the FORT can be maintained but few of the atoms will absorb photons. The important advantages of FORTs compared to MOTs are: (1) high density ($\sim 10^{12}$ cm^{-3}) and (2) a well-defined polarization axis along which atoms can be aligned or oriented (spin-polarized). The main disadvantage is the small number of trapped atoms due to small FORT volume. The best number achieved is about 10^4 atoms.

6.4.5 Magnetic traps

Pure magnetic traps have also been used to study cold collisions, and they are critical for the study of dilute gas-phase Bose–Einstein condensates (BECs) in which collisions figure importantly. We anticipate therefore that magnetic traps will play an increasingly important role in future collision studies in and near BEC conditions.

The most important distinguishing feature of all magnetic traps is that they do not require light to provide atom containment. Light-free traps reduce the rate of atom heating by photon absorption to zero, an apparently necessary condition for the attainment of BEC. Magnetic traps rely on the interaction of atomic spin with variously shaped magnetic fields and gradients to contain atoms. The two governing equations are

$$U = -\mu_{\mathbf{S}} \cdot \mathbf{B} = -\frac{g_s \mu_B}{\hbar} \mathbf{S} \cdot \mathbf{B} = -\frac{g_s \mu_B}{\hbar} M_S B \tag{6.40}$$

and

$$\mathbf{F} = -\frac{g_s \mu_B}{\hbar} M_S \nabla B \tag{6.41}$$

If the atom has nonzero nuclear spin I then $\mathbf{F} = \mathbf{S} + \mathbf{I}$ substitutes for $\mathbf{S}$ in Eq. 6.40, the g-factor generalizes to

$$g_F \cong g_S \frac{F(F+1) + S(S+1) - I(I+1)}{2F(F+1)} \tag{6.42}$$

and

$$\mathbf{F} = -\frac{g_F \mu_B}{\hbar} M_F \nabla B \tag{6.43}$$

Depending on the sign of U and $\mathbf{F}$, atoms in states whose energy increases or decreases with magnetic field are called "weak-field seekers" or "strong-field seekers," respectively. One could, in principle, trap atoms in any of these states, needing only to produce a minimum or a maximum in the magnetic field. Unfortunately only weak-field seekers can be trapped in a static magnetic field because such a field in free space can only have a minimum. Even when weak-field-seeking states are not in the lowest hyperfine levels, they can still be used for trapping because the transition rate for spontaneous magnetic dipole emission is $\sim 10^{-10}\,\mathrm{s}^{-1}$. However, spin-changing collisions can limit the maximum attainable density.

The first static magnetic field trap for neutral atoms used an anti-Helmholtz configuration, similar to a MOT, to produce an axially symmetric quadrupole magnetic field. Since this field design always has a central point of vanishing magnetic field, nonadiabatic Majorana transitions can take place as the atom passes through the zero point, transferring the population from a weak-field to a strong-field seeker and effectively ejecting the atom from the trap. This problem can be overcome by using a magnetic bottle with no point of zero field. The magnetic bottle, also called the *Ioffe-Pritchard trap*, has been used to achieve BEC in a sample of Na atoms precooled in a MOT. Other approaches to eliminating the zero field point are the time-averaged orbiting potential (TOP) trap and an optical "plug" that consists of a blue-detuned intense optical beam aligned along the magnetic trap symmetry axis and producing a repulsive potential to prevent atoms from entering the null field region. Trap technology continues to develop and BEC studies will stimulate the search for more robust traps containing greater numbers of atoms. At present $\sim 10^7$ atoms can be trapped in a Bose–Einstein condensate loaded from a MOT containing $\sim 10^9$ atoms.

6.5 Further Reading

The optical dipole gradient force and the "radiation pressure" force are discussed in many books. A clear and careful discussion can be found in

- C. Cohen-Tannoudji, J. Dupont-Roc, and G. Grynberg, *Atom-Photon Interactions*, chapter V, Wiley-Interscience, New York, 1992.

A semiclassical development of the radiation pressure force and laser cooling can be found in

- S. Stenholm, *Rev. Mod. Phys.* **58**, 699–739 (1986).

An interesting discussion of the dipole gradient force and the radiation pressure force presented as aspects of the classical Lorentz force acting on a harmonically bound electron can be found in

- S. C. Zilio and V. S. Bagnato, *Am. J. Phys.* **57**, 471–474 (1989).

We have followed the 1-D sub-Doppler cooling models developed in

- J. Dalibard and C. Cohen-Tannoudji, *J. Opt. Soc. Am. B* **6**, 2023–2045 (1989).

A more detailed and quantitative discussion of these models with comparison to experiment can be found in

- P. D. Lett, W. D. Phillips, S. I. Rolston, , C. I. Tanner, R. N. Watts, and C. I. Westbrook, *J. Opt. Soc. Am. B* **6**, 2084–2107 (1989).

An excellent review of cooling and trapping neutral atoms, including a detailed discussion of magnetic trapping can be found in,

- H. Metcalf, and P. van der Straten, Physics Reports **244**, 203–286 (1994).

Further Reading

Chapter 7

The Laser

7.1 Introduction

Once derided as "a solution looking for problems," the laser has come into its own. In daily life, we depend on lasers in telecommunications, in medicine, in audio and video entertainment, at the checkout counter at the supermarket. In the laboratory, the laser has revolutionized many fields of research from atomic and molecular physics to biology and engineering. Just as varied as the applications of lasers are their properties. The physical dimensions vary from 100 μm (semiconductor lasers) to the football stadium size of the Nova laser at the Lawrence Livermore National Laboratory in Livermore, California. The average output power ranges from 100 μW to kilowatt levels. The peak power can be as high as 10^{15} W, and the pulse duration can be as short as 10^{-14} s. Wavelengths range from the infrared to the ultraviolet. Yet behind the infinite variety, the operating principle of all lasers is essentially the same. The laser is an oscillator working in the optical region. Like the electronic oscillator, the laser consists of two main components: a gain medium and a resonator. The *gain medium* consists of excited atoms that amplify the signal, namely, the optical field, by stimulated emission, as described in Chapters 1 and 2. Gain is obtained when the atoms are, on average, excited to the upper level more than the lower level, so that the excess energy between the two levels can be given to the optical field by stimulated emission. The *resonator* provides frequency-selective positive feedback that feeds part of the amplified field back to the gain medium repeatedly. Part of the field exits the resonator as output of the laser. The field inside the resonator cannot be amplified indefinitely, because by conservation of energy, the amplification of the gain medium has to decrease when the field is high enough; that is, the gain saturates. Free running, the laser reaches a steady state when the saturated gain is equal to, and compensates for, the loss of the field through output and other possible causes such as absorption in the resonator components.

The most distinctive feature of the laser is called *coherence*, which refers to the degree to which one can predict the field of the beam. If, with the knowledge of the optical field at one point, one can predict the field at a later time at the same place, the beam is said to be temporally coherent. In this sense, a purely monochromatic field is perfectly coherent. Laser beams are highly monochromatic, with a frequency width that has been narrowed to as little as 10^{-15} of the center frequency. Monochromaticity is a result of the extreme frequency discrimination afforded by the combination of stimulated emission and resonator feedback. The field emitted by stimulated emission bears a definite phase relationship to the field that stimulates the emission. Because the field inside the laser resonator builds up by repeated stimulated emission, the final field reached is in phase with the initial starting field. However, the initial starting field comes from spontaneous emission, whose frequency can spread over a wide range. Because of the frequency-selective feedback of the resonator, the one frequency at the peak of the resonance with the highest feedback extracts the most energy from the gain that it saturates. The saturated gain is exactly equal to the loss at the peak of the resonance. Frequencies that deviate even slightly from the resonance will see a net loss and the field cannot build up. The final field, however, is subject to many noisy perturbations such as thermal fluctuations of various kinds, vibrations, and spontaneous emissions from the gain medium. These perturbations ultimately add noise to the laser field and are responsible for the small but finite frequency width of the laser field. The fundamental limit to the finite frequency width comes from spontaneous emission, which cannot be entirely eliminated from the lasing transition. The spontaneously emitted field, which is random in nature, can be either in phase with the laser field, or in quadrature. The in-phase spontaneous emission affects the amplitude of the laser field and is suppressed by gain saturation. The in-quadrature component of spontaneous emission changes only the phase, not the amplitude, of the laser field, and is therefore not suppressed by the saturated gain . It is this random phase fluctuation that gives the laser field its finite frequency width. It is important to recognize that in a laser, the optical field and the atoms are inseparably coupled by the feedback of the resonator. In fact, all the atoms participating in the lasing action interact with the common optical field and oscillate together as a giant dipole (called *macroscopic polarization*). It is this collective action that results in a frequency width much narrower than the "natural" width allowed in uncoupled atoms, despite the constant dephasing and decay processes in a laser.

The laser beam is also spatially coherent; we can predict the field at another place with the knowledge of it at one place. Spatial coherence is a direct result of the resonator. An optical resonator can be designed so that one spatial mode suffers less loss than any other. The spatial mode with the least loss (or highest Q) ultimately oscillates, whereas the others are suppressed, again by the mechanism of gain saturation.

In this chapter, we will adapt the fundamental equations governing a two-level system, developed in Chapters 2 and 4, to the laser oscillator. The difference between the previous situation of an isolated two-level atom and the present one of the laser oscillator is in the electromagnetic field. In Chapter

2, we considered the electromagnetic wave only at the point where it interacts with the atom. The displacement of the electron under the electric field is small compared to the wavelength. In a laser oscillator, the electromagnetic field is confined in an optical resonator, and is a standing wave. The volume of interaction extends over many wavelengths. We will consider the homogeneously broadened medium in detail, and only single mode operation. Multimode operation and inhomogeneous broadening are discussed only briefly.

7.2 Single-Mode Rate Equations

The optical Bloch equations for a two-level atom (Chapter 4) will be developed further. These equations will be summed statistically over all the atoms interacting with the optical field in the resonator, and damping terms will be introduced phenomenologically. After this operation, elements of the density matrix in these equations become macroscopic physical quantities. The diagonal elements become the number of atoms per unit volume in the lower and upper states, and the off-diagonal elements, multiplied by the interaction matrix element $\boldsymbol{\mu}$, become the macroscopic polarization $\mathbf{P}$. The result is a set of coupled equations for the population densities and polarization. To complete the description, one more equation for the optical field is derived directly from the classical Maxwell equations. The complete set of equations describe the motion of three physical quantities: population inversion, polarization, and the electric field. To reduce the number of equations, the polarization is integrated and expressed in terms of the other two quantities under the assumption that the populations vary slowly compared to the polarization. The resulting two equations describing the dynamics of population inversion and light intensity are called the *rate equations*.

Optical fields in a resonator are discussed in detail in Chapter 8. They are counterpropagating Gaussian beams. The exact spatial distribution of the Gaussian beams is not needed here. Instead, we will approximate the beam inside the resonator with a plane standing wave.

It is fruitful to approach the rate equations from a classical point of view. Quantum mechanics is used solely to describe the atoms leading to the macroscopic polarization and the gain. In Appendix 7.A, the macroscopic polarization from a group of classical harmonic oscillators is first derived, then by physical argument, converted into the polarization of a group of atoms. In the process, several physical quantities are introduced heuristically: the classical electron radius, the classical radiative lifetime, the quantum radiative lifetime, and the interaction cross section. The cross section deserves special attention. We have already introduced the absorption and emission cross sections in our discussion of the two-level atom (see Chapter 2, Section 2.2.6). The strength of interaction between the optical field and the atoms is given by the interaction matrix element $\boldsymbol{\mu}$ (see Chapter 2, Section 2.2.1). Instead of $\boldsymbol{\mu}$, the equivalent, but perhaps physically more appealing quantity, the interaction cross section σ can be used. This is more than just a change of notation. The concept of cross section appears throughout physics, chemistry, and engineering to quantify the strength of an interaction or process. In fact, in many calculations involving lasers, only the relevant cross sections and the population decay times are needed.

It is also interesting to examine the classical electronic oscillator whose amplification is provided by a negative resistance. This is carried out in Appendix 7.B, where the equations for the voltage amplitude and phase are shown to be identical in form to those derived for the laser field amplitude and phase. By introducing an external voltage source, which can represent either an injecting signal, or noise, two important subjects can be conveniently discussed: injection locking and phase noise that ultimately limits monochromaticity in an oscillator

7.2.1 Population inversion

The mathematical description of atom–field interaction using the optical Bloch equations has been developed in Chapter 4. The density matrix equations (Eqs. 4.5) are reproduced below:

$$\frac{d\rho_{21}}{dt} = -i\frac{\mu E_0}{2\hbar}(-\rho_{22}+\rho_{11})e^{-i(\omega-\omega_0)t} = \frac{d\rho_{12}^*}{dt} \tag{7.1}$$

$$\frac{d\rho_{22}}{dt} = i\frac{\mu E_0}{2\hbar}(\rho_{21}e^{i(\omega-\omega_0)t} - \rho_{12}e^{-i(\omega-\omega_0)t}) = -\frac{d\rho_{11}}{dt} \tag{7.2}$$

The factor appearing in these two equations, $\mu E_0/\hbar$, called the *Rabi frequency*, quantifies the rate by which the density matrix elements change under the electric field. In fact, instead of the optical Bloch equations, if the equations for wavefunction coefficients C_1 and C_2 are integrated directly, it can be found that $|C_1^2|$ and $|C_2^2|$ oscillate with a frequency $\mu E_0/\hbar$ (Chapter 2, Section 2.2.2 and Chapter 4, Section 4.4). The diagonal and off-diagonal elements are coupled via the electric field. We will first eliminate the off-diagonal elements.

Equations 7.1, and 7.2 are now to be considered as *statistically averaged over the atoms.* The statistically averaged ρ_{21} describes the phase correlation or *coherence* between the two eigenstates of the atom, but this correlation can be degraded. As in the treatment for the individual two-level atom (Section 4.5), this degradation is treated phenomenologically here by a dephasing rate constant Γ. Equation 7.1 then becomes

$$\frac{d\rho_{21}}{dt} + \Gamma\rho_{21} = i\frac{\mu E_0}{2\hbar}(\rho_{22}-\rho_{11})e^{-i(\omega-\omega_0)t} \tag{7.3}$$

where the matrix elements are understood to have been statistically averaged. Dephasing can be caused by many processes: collision, decay of populations including spontaneous emission[1], interaction with surrounding host molecules. Since it includes decay, Γ is at least as large as the population rate of change (including both decay and changes induced by the field); often, in fact, it is much

[1]Note that Γ here, when identified with spontaneous emission, is equivalent to the γ in Section 4.5

larger. In such cases, we can treat the populations as constant and integrate Eq. 7.3 to obtain[2]

$$\rho_{21} = -\frac{\mu E_0}{2\hbar} \frac{\rho_{22} - \rho_{11}}{(\omega - \omega_0) + i\Gamma} e^{-i(\omega-\omega_0)t} \tag{7.4}$$

where ρ_{21} depends on the difference $\rho_{22} - \rho_{11}$, not ρ_{22} or ρ_{11} individually. To amplify the optical field, the atoms must be, on average, excited more to the upper level than the lower level: where $\rho_{22} > \rho_{11}$. This condition is called *population inversion*. The gain is proportional to the population difference; it is therefore desirable, and often achievable, to have $\rho_{22} >> \rho_{11}$. Then $\rho_{22} - \rho_{11} \backsimeq \rho_{22}$. This is realized in many lasers, whose gain media and levels are chosen such that the decay rate of the lower level is much faster than that of the upper level, and the rate of excitation to the upper level much greater than to the lower level[3]. Unless the intensity of light in the medium is so strong that the lower level is significantly populated by stimulated emission, we can approximate the population inversion, $N_0(\rho_{22} - \rho_{11}) \equiv \Delta N$ by $N_0\rho_{22}$, where N_0 is the total atomic density.[4] Statistically averaging Eq. 7.2, we have

$$\frac{d}{dt}\Delta N + \frac{\Delta N}{T_1} = R_{\text{pump}} + i\frac{\mu E_0}{2\hbar} N_0(\rho_{21} e^{i(\omega-\omega_0)t} - \rho_{12} e^{-i(\omega-\omega_0)t}) \tag{7.5}$$

where T_1 is the decay time constant of ΔN. An additional term R_{pump} was introduced on the right-hand side to represent *pumping*. Pumping is the excitation process whereby the atoms are excited to the upper level. There are many methods of pumping, each appropriate to a particular laser system. Common pumping methods include optical excitation by either coherent or incoherent sources, electric currents, and discharge. In general, lasers are inefficient devices, with under 1% of the total input energy converted into light. There are two reasons. The main reason is the inefficiency of the pumping process. For example, in flashlamp pumping, most of the lamp light is not absorbed by the atoms because of its broad spectrum; in fact, most of the lamp light does not even reach the atoms because it is difficult to focus the spatially incoherent lamp light. The second reason is the energy levels of the atoms. It is impossible to achieve population inversion by interaction with only two levels of the atom—at least one more level must be involved. The photon energy, which is equal to the energy difference between the two levels in the stimulated emission process, is less than the energy difference between the highest and lowest energy levels involved in the whole pumping process. The ratio of the energy of the stimulated emission transition to the energy of the pumping transition is called the *quantum efficiency*, which can be smaller than 0.1. We substitute ρ_{21} from

[2]We emphasize the importance of not confusing Γ in Eq. 7.4 with use of Γ for the rate of spontaneous emission (see Chapter 4, Eq. 4.19).

[3]One exception is the erbium-doped fiber, discussed below.

[4]The use of N_0 and ΔN for atomic density here differs slightly from the notation N/V used in Eqs. 4.16 and 4.22 and subsequent expressions for the susceptibility in Chapter 4.

Eq. 7.4 into Eq. 7.5 to obtain

$$\frac{d}{dt}\Delta N + \frac{\Delta N}{T_1} = R_{\text{pump}} - \frac{1}{2}\left(\frac{\mu E_0}{\hbar}\right)^2 \Delta N \frac{\Gamma}{(\omega - \omega_0)^2 + \Gamma^2} \tag{7.6}$$

We will cast the last term on the right-hand side into another form. To do that, let us digress and revisit the macroscopic polarization. The expectation value of the transition dipole moment d of one atom is

$$d \equiv \int \Psi^*(-ex)\Psi d^3x = -\mu(\rho_{21}e^{-i\omega_0 t} + \rho_{12}e^{i\omega_0 t})$$

Let P be the component of the polarization $\mathbf{P}$ in the direction of the electric field that induces the dipole moments. Then $P = N_0 \langle d \rangle$, where the angular brackets denote statistical average

$$P = -N_0\mu(\rho_{21}e^{-i\omega_0 t} + \rho_{12}e^{i\omega_0 t}) \tag{7.7}$$

now where it is understood that the density matrix elements have been statistically averaged.[5] It is significant that the average over many atoms does not vanish. It means that atoms are radiating in synchronism. The synchronism is established by the common driving electric field. The electric field, in turn, is produced by the collective radiation from the atoms. Another manifestation of the collective radiation is the generation of the laser beam; the radiation pattern of a single dipole is donut-shaped—the beam is the result of coherent superposition of many dipole fields.

The polarization P is a real quantity. We define a *complex* polarization $\mathcal{P}$

$$\mathcal{P} = -2N_0\mu\rho_{21}e^{-i\omega_0 t} \tag{7.8}$$

so that

$$P = \frac{1}{2}(\mathcal{P} + \mathcal{P}^*)$$

The complex polarization $\mathcal{P}$ can be calculated by substituting ρ_{21} from Eq. 7.4:

$$\begin{aligned} \mathcal{P} &= \frac{\Delta N \mu^2}{\hbar} \frac{E_o e^{-i\omega t}}{(\omega - \omega_o) + i\Gamma} \\ &\equiv \epsilon_o \chi \, E_o e^{-i\omega t} \end{aligned} \tag{7.9}$$

where we have, again, the susceptibility[6]

$$\chi(\omega) = \frac{\mu^2}{\epsilon_0 \hbar} \frac{\Delta N}{(\omega - \omega_o) + i\Gamma} = \frac{\mu^2}{\epsilon_0 \hbar} \Delta N \frac{(\omega - \omega_o) - i\Gamma}{(\omega - \omega_o)^2 + \Gamma^2} \tag{7.10}$$

$$\equiv \chi' + i\chi'' \tag{7.11}$$

[5]Note that Eq. 7.7 is the macroscopic analog of Eqs. 4.16 and 4.17 and that Eqs. 4.3 enable the identification of ρ_{12} with $C_1^* C_2$ and ρ_{21} with $C_2 C_1^*$ in the two-level atom.

[6]Note the difference of a factor of 2 in the Γ term between Eqs. 7.10 and 4.27.

Going back to Eq. 7.6, we can substitute μ^2 and the Lorentzian by χ'' and obtain

$$\frac{d\Delta N}{dt} + \frac{\Delta N}{T_1} = R_{\text{pump}} + \frac{\chi''}{c\hbar}\left(\frac{c\epsilon_0}{2}E_0^2\right) \tag{7.12}$$

Note that

$$\chi'' = -\frac{\Delta N \mu^2}{\epsilon_0 \hbar}\frac{\Gamma}{(\omega-\omega_0)^2+\Gamma^2} \tag{7.13}$$

is negative for positive ΔN. The power exchange between the field and the atoms is via χ'', the imaginary part of the susceptibility. It means that it is the in-quadrature component of the dipole that accounts for the power exchange with the field (see Section 4.4.2). The exchange comes from the familiar expression for power, force times velocity. The force is electron charge times the electric field. If the power is to be nonzero after averaging over one cycle, then the velocity must be in phase with the field. Since the velocity is the time derivative of electron displacement, the two quantities are in quadrature, which means that the displacement is in quadrature with the field. Since the dipole moment is simply electron charge multiplied by displacement, the last term on the right-hand side of Eq. 7.12 describes the rate of change of the population inversion due to stimulated emission. As discussed in Section 1.1.2 (see Eq. 1.10), the factor in brackets, $(c\epsilon_0/2)\,E_0^2$, is the light intensity I averaged over one cycle. We rewrite that term as

$$\frac{\chi''}{c\hbar}\left(\frac{c\epsilon_0}{2}E_0^2\right) = -\sigma(\omega)\Delta N\frac{I}{\hbar\omega} \tag{7.14}$$

where $\sigma(\omega)$ is the *transition cross section* [7]

$$\sigma(\omega) = -\frac{\omega}{c}\chi''(\omega)\frac{1}{\Delta N} \tag{7.15}$$

The concept of interaction cross section is an important one. It has been already discussed in Section 2.2.6 and will be further discussed in Appendix 7.A. It is a fictitious area that characterizes the strength of the interaction in question. The radiative transition rate, the left-hand side of Eq. 7.14, is equal to a product of three terms: the cross section, the population inversion, and the photon flux $I/\hbar\omega$. From Eqs. 7.12– 7.14 it follows that the cross section can be written in terms of a "peak cross section" and a unitless lineshape form factor. The peak cross section is

$$\sigma_0 = \frac{\omega\mu^2}{c\epsilon_0\hbar}\cdot\frac{1}{\Gamma}$$

and the form factor expressing the spectral width of the cross section in terms of the "peak" value is

$$\frac{\Gamma^2}{(\omega-\omega_0)^2+\Gamma^2} \tag{7.16}$$

[7] Compare Eqs. 7.15 and 4.31 for the absorption cross section in terms of the susceptibility in a two-level atom.

such that

$$\sigma(\omega) = \frac{\sigma_0}{1 + (\omega - \omega_0)^2/\Gamma^2} \tag{7.17}$$

If we now substitute the expression for μ^2 in terms of the spontaneous emission rate (Eq. 2.22), we can write the peak cross section as

$$\sigma_0 = \frac{3\lambda_0^2}{2\pi} \cdot \frac{A_{21}}{2\Gamma} \tag{7.18}$$

Note that if the only source of dissipation is spontaneous emission, the 2Γ in the denominator of Eq. 7.18 is just A_{21}, and the peak cross section is simply

$$\sigma_0 = \frac{3\lambda_0^2}{2\pi} \tag{7.19}$$

It is worth pointing out that $\sigma(\omega)$ in Eq. 7.17 is not the same as the "spectral" cross section (see Eq. 2.32)

$$\sigma_\omega = \frac{g_2}{g_1}\frac{\lambda^2}{4} \cdot A_{21}$$

which has units of the product of area and frequency and is the relevant expression from which integration over the Lorentzian line shape $F(\omega - \omega_0)$ yields a cross section in units of area. As discussed in Chapter 2, Section 2.2.6, the result of this integration over the "natural" line width is

$$\sigma_{0a} = \frac{g_2}{g_1}\frac{\pi\lambda_0^2}{2} \tag{7.20}$$

There are two reasons why σ_0 and σ_{0a} are not identical: (1) Eq. 7.19 does not include averaging over random orientations of the transition dipole while Eq. 7.20 does. The effect of this averaging is just to multiply Eq. 7.19 by a factor of $\frac{1}{3}$; and (2) Eq. 7.19 is the cross section at line center, while Eq. 7.20 is integrated over the whole line shape. It is not difficult to show that these two cross section expressions are consistent by converting the peak cross section (Eq. 7.19) to a "spectral gradient cross section" by dividing the peak by the spectral width and then integrating it over the form factor, Eq. 7.16. The result is equivalent to Eq. 7.20.

It is remarkable that the intrinsic cross section of *all* dipole transitions are, within a numerical factor on the order of unity, equal to the wavelength squared. In deriving the expressions for the peak cross section (Eqs. 7.18, 7.19) and its frequency dependence (Eq. 7.17) we have assumed that the electric field and the transition dipole are aligned along the axis of quantization. In many cases, however, the atoms' transition dipoles are randomly oriented, and as much as two-thirds of them can be aligned relative to the electric field in such a way that their dipole moments are zero. In that case, the average peak cross section is only one third of the expressions given above. We follow Professor Siegman's notation (A. E. Siegman, *Lasers*, University Science Book, 1986) and replace the factor 3 by 3^*, whose value can range from 1 to 3. Furthermore, we have arrived at these expressions assuming that the atoms are in vacuum. If they are in a material of refractive index n, then we must divide the wavelength by n. The final expression for the peak cross section is then

$$\sigma_0 = \frac{3^*}{2\pi}\frac{\lambda^2}{n^2}\left(\frac{A_{21}}{2\Gamma}\right) \tag{7.21}$$

Equation 7.12, now rewritten in the final form below, is one of a pair of equations commonly known as the laser *rate equations*:

$$\frac{d\Delta N}{dt} + \frac{\Delta N}{T_1} = R_{\text{pump}} - \sigma(\omega)\Delta N \frac{I}{\hbar\omega} \tag{7.22}$$

7.2.2 Field equation

The fields in an optical resonator are fields of the *modes* of the resonator. The field of a mode has a definite spatial pattern whose amplitude oscillates in time at the mode frequency. When an atomic medium is introduced into the otherwise empty resonator, the modes are changed in two ways. The permittivity of the medium changes the phase velocity of light and therefore changes the mode frequency and the wavelength inside the medium. The lasing transition provides further changes; with population inversion, amplification is possible to compensate for the loss in the resonator, and the susceptibility of the lasing transition changes the mode frequency so that the final oscillation frequency must be determined taking the dynamics of the lasing action into account. Most laser resonators are formed of spherical mirrors, and the fields inside are counterpropagating Gaussian beams, the subject Chapter 8. Near the axis of the Gaussian beam, however, the fields very closely resemble plane waves. In this chapter, we will use two counter-propagating plane waves for the field, with the boundary condition that they vanish at the mirrors. This boundary condition results in a standing wave and simplifies the mathematics. Light must exit from one of the mirrors to provide an output, and it may be attenuated in the resonator by absorption or scattering. These losses are accounted for by a fictitious, distributed loss instead, again for mathematical simplicity.

The second of the rate equations is derived directly from the Maxwell equations, two of which are reproduced here,

$$\begin{aligned} \nabla \times \mathbf{E} &= -\mu_0 \frac{\partial \mathbf{H}}{\partial t} \\ \nabla \times \mathbf{H} &= \mathbf{J} + \frac{\partial \mathbf{D}}{\partial t} \end{aligned}$$

The equations say that a time varying magnetic field acts as a source for the electric field and vice versa. The displacement field $\mathbf{D}$ consists of two parts, the electric field and the macroscopic polarization (see Section 1.4):

$$\mathbf{D} = \epsilon_0 \mathbf{E} + \mathbf{P}$$

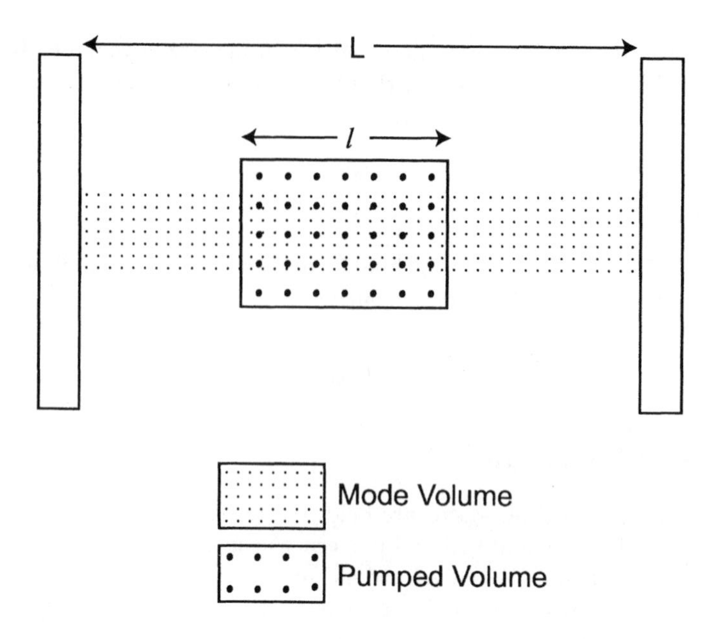

Figure 7.1: The laser resonator with gain medium . The resonator mode volume and the gain overlap. The ratio of the overlapping gain volume to the mode volume is the filling factor s.

Normally in a laser gain medium, there is no conduction current varying at the optical frequency. To account for the loss of the field in the resonator, we introduce a fictitious conductivity σ_c so that

$$\mathbf{J} = \sigma_c \mathbf{E}$$

We will see how to relate this fictitious conductivity to real losses later.

Eliminating the magnetic field from the two Maxwell's equations, assuming little transverse variation of the fields, and approximating

$$\frac{\partial^2 \mathbf{P}}{\partial^2 t} \to -\omega^2 \mathbf{P}$$

we obtain the wave equation for the electric field:

$$-\nabla^2 E + \mu_0 \sigma_c \frac{\partial E}{\partial t} + \frac{1}{c^2}\frac{\partial^2 E}{\partial t^2} = \mu_0 \omega^2 P \tag{7.23}$$

We have assumed that E and P are polarized in only one direction (x, say), and vary only in the propagation direction (z). We now use a simple model of the resonator. It consists of a pair of perfectly reflecting mirrors located at $z = 0$ and $z = L$ where the total electric field vanishes (Fig. 7.1). We write E in the form

$$E(z,t) = U(z)\left[\frac{E_0(t)}{2}e^{-i\omega t - i\phi(t)} + \frac{E_0(t)}{2}e^{i\omega t + i\phi(t)}\right]$$

where

$$U(z) = \sin(Kz)$$

with $K = n\pi/L$ to satisfy the boundary conditions at the mirrors, and n is an integer. The boundary conditions yield the mode resonance frequency

$$\Omega = Kc = 2n\pi\frac{c}{2L}$$

There are infinitely many modes, with their angular frequencies separated by a multiple of $c\pi/L$. The interval between two adjacent modes is usually smaller than Γ, the transition width. The mode with frequency closest to the peak of the gain, at ω_0, extracts the most energy from the medium and we assume that this mode alone oscillates. The amplitude $E_0(t)$ has been chosen to be real, and a time-varying phase $\phi(t)$ is allowed.

The pumped atomic medium overlaps with the mode volume. In the overlapping region, the medium partially fills a fraction s of the mode volume (Fig. 7.1). It is described by the polarization, which, in terms of susceptibility χ, is

$$P(z,t) = V(z)\epsilon_0\left[\chi(\omega)\frac{E_0(t)}{2}e^{-i\omega t-i\phi(t)} + \text{c.c.}\right]$$

The spatially varying function $V(z)$ is equal to $U(z)$ inside the medium and zero outside. We now take the time derivatives of E and P, substitute into Eq. 7.23, multiply the whole equation by $U(z)$, and integrate over z from 0 to L. The fields are almost monochromatic; therefore in the conduction current term $\partial E/\partial t \sim -i\omega E$. In the term $\partial^2 E/\partial^2 t$, however, the leading term -$\omega^2 E$ is almost completely canceled by the spatial derivative term

$$\nabla^2 E = -\frac{\partial^2 E}{\partial z^2} = -K^2 E$$

therefore the next order term

$$-2i\omega\left[\frac{dE_0}{dt} - iE_0\frac{d\phi}{dt}\right]U(z)$$

must be kept. After these operations, and after separating and equating the real and imaginary parts, Eq. 7.23 becomes two equations, one for the amplitude, the other for the phase, of the electric field:

$$\frac{dE_0}{dt} + \frac{\sigma_c}{2\epsilon_0}E_0 = -s\frac{\omega}{2}\chi'' E_0 \tag{7.24}$$

$$\frac{d\phi}{dt} + (\omega - \Omega) = -s\frac{\omega}{2}\chi' \tag{7.25}$$

Equation 7.24 says that the field amplitude decays through the conductivity and grows by $-\chi''$, the growth rate of the *field* is $-\omega\chi''/2$, reduced by the filling factor s of the resonator mode volume. The growth rate of the *intensity* I is twice that of the field. Multiplying the whole Eq. 7.24 by E_0 converts it into an intensity equation:

$$\frac{dI}{dt} + \frac{\sigma_c}{\epsilon_0} I = -s\omega\chi'' I \tag{7.26}$$

The right-hand side of Eq. 7.26 is the rate of growth of light intensity. The energy comes from the medium. Using Eq. 7.15 to replace χ'' with the cross section σ, we rewrite Eq. 7.26 as

$$\frac{dI}{dt} + \frac{\sigma_c}{\epsilon_0} I = sc\sigma\Delta N I \tag{7.27}$$

Equation 7.27 is the second of the rate equations. The quantity $\sigma\Delta N$ is the optical *gain* per unit length, and $c\sigma\Delta N$ is the optical gain per unit time, or growth rate. This equation can be written in a slightly different form. Dividing the equation by c and replacing cdt by dz, we have

$$\frac{dI}{dz} + \frac{\sigma_c}{\epsilon_0 c} I = s\sigma\Delta N I \tag{7.28}$$

This equation describes the propagation of a traveling wave through a gain medium and would have been obtained had we started with a traveling wave instead of a standing wave.

Equation 7.25 determines the oscillation frequency ω. The real part of the susceptibility χ' is zero only at the transition line center ω_0. If the resonator is not tuned to ω_0, then ω must be determined from Eq. 7.25; it will be neither Ω nor ω_0 but a value between the two.

Finally, to relate the fictitious conductivity to resonator loss, consider a resonator whose mirror reflectivities are R_1 and R_2, and the loss in the atomic medium is α per unit length along its length l. Integrating Eq. 7.28 without gain for one-round trip time T_R inside the resonator, we have

$$I(T_R) = I(0)e^{-(\sigma_c/\epsilon_0)T_R}$$

Now, if we follow the light in one round trip, the fraction returned is $R_1R_2e^{-2\alpha l}$. So

$$e^{-(\sigma_c/\epsilon_0)T_R} = R_1R_2e^{-2\alpha l} \tag{7.29}$$

7.3 Steady-State Solution to the Rate Equations

Solving the rate equations in the steady state leads us to several important concepts, and gives us the most important information on the laser. The concepts introduced are: small-signal or unsaturated gain, saturated gain, oscillation or lasing threshold, and saturation intensity. In terms of the saturation intensity and the degree that the unsaturated gain is above threshold, the laser intensity can be immediately calculated. Then, the phase equation will be examined to see the effect of *frequency pulling* and *pushing* on the oscillation frequency. Although the solution is formally for the steady state (i.e., all quantities do not vary in time), it is also valid when the laser operates in a pulsed mode, if the pulse is longer than the population decay time T_1.

With all time derivatives set to zero, the rate equations, Eqs. 7.22 and 7.27 become

$$\frac{\Delta N}{T_1} = R_{\text{pump}} - \sigma \Delta N \frac{I}{\hbar\omega} \tag{7.30}$$

$$\frac{\sigma_c}{\epsilon_0} I = sc\sigma \Delta N I \tag{7.31}$$

The population inversion ΔN can be found from Eq. 7.30 in terms of I

$$\begin{aligned} \Delta N &= \frac{R_{\text{pump}} T_1}{1 + I/I_S} \\ &\equiv \frac{\Delta N_0}{1 + I/I_S} \end{aligned} \tag{7.32}$$

where we have introduced the saturation intensity

$$I_S = \frac{\hbar\omega}{\sigma T_1} \tag{7.33}$$

and the small-signal population inversion

$$\Delta N_0 = R_{\text{pump}} T_1$$

which is the population inversion when there is no light ($I = 0$) or when the light is weak ($I << I_S$). The optical gain (per unit length), in the presence of light, called *saturated gain*, is

$$\sigma \Delta N = \frac{\sigma \Delta N_0}{1 + I/I_S} \tag{7.34}$$

The numerator $\sigma \Delta N_0$ on the right-hand side is the gain (per unit length) when there is no light ($I = 0$) . It is called the *small-signal* or *unsaturated gain*. The saturated gain is smaller than the unsaturated gain because the population inversion has been lowered by stimulated emission to provide energy to the light.

Equation 7.31 allows two solutions, below and above *threshold*. We can imagine operating a laser by gradually increasing the pumping (ΔN_0). When the pumping is low, so is the gain $sc\sigma \Delta N_0$ (the right-hand side of Eq. 7.31) and is less than σ_c/ϵ_0 (the left-hand side of Eq. 7.31). To satisfy the equation, I must be zero: the gain is not enough to overcome the loss and the laser is not operational. The laser is said to be below *threshold*. The threshold is reached when $sc\sigma \Delta N_0$ is equal to σ_c/ϵ_0. We define the threshold population inversion ΔN_{th} such that

$$sc\sigma \Delta N_{\text{th}} = \frac{\sigma_c}{\epsilon_0} \tag{7.35}$$

Increasing pumping further increases the small-signal gain; but the saturated gain, pulled down by the now non-zero intensity, remains constant and equal to the loss σ_c/ϵ_0:

$$sc\sigma \Delta N = \frac{\sigma_c}{\epsilon_0}$$

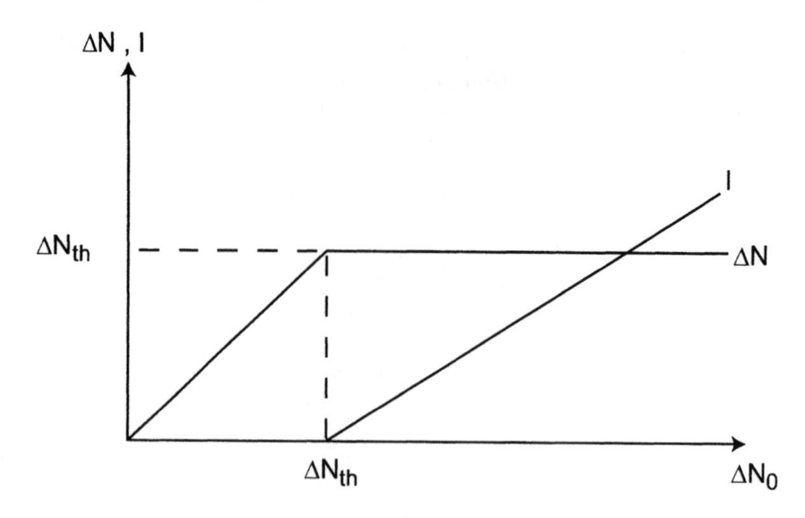

Figure 7.2: Population inversion ΔN and light intensity I versus ΔN_0. Threshold is defined as ΔN_{th}. Below threshold $\Delta N = \Delta N_0$, and $I = 0$. Above threshold, $\Delta N = \Delta N_{\text{th}}$ and $I = I_s \left[\frac{\Delta N_0}{\Delta N_{\text{th}}} - 1\right]$

or

$$\Delta N = \Delta N_{\text{th}}$$

when $\Delta N_0 \geq \Delta N_{\text{th}}$.

Substitution ΔN from Eq. 7.34 into Eqs. 7.30 and7.31 yields the intensity above threshold:

$$I = I_S \left[\frac{\Delta N_0}{\Delta N_{\text{th}}} - 1\right] \tag{7.36}$$

This is an important result. In words, the light intensity *inside* a laser is equal to the saturation intensity of the lasing transition, multiplied by the fraction that the pumping is above threshold.

Figure 7.2 shows ΔN and I versus pumping as represented by ΔN_0. Below threshold $I = 0$, and the saturated gain is equal to the unsaturated gain. Above threshold, I rises linearly with pumping, while the saturated gain is pinned to the loss.

Suppose one mirror of the resonator is perfectly reflecting, $R_1 = 1$, the other transmits a fraction $T = 1 - R_2$ of the incident intensity. The cross-sectional area of the beam is determined by the resonator design, discussed in Chapter 8, but suppose that it is A. The light intensity consists of two equal counterpropagating parts, one toward the transmitting mirror. Then the *output power* of the laser is

$$\begin{aligned} P_{\text{out}} &= \frac{1}{2} TAI \\ &= \frac{1}{2} TAI_S \left[\frac{\Delta N_0}{\Delta N_{\text{th}}} - 1\right] \end{aligned} \tag{7.37}$$

In practice, pumping usually cannot be too close to, or too much above, threshold. The laser tends to be unstable when it is too close to threshold. When it is well above threshold, many problems, including heating and undesirable nonlinear effects, can occur. *It is therefore a useful rule of thumb that the intensity of the laser beam inside the resonator is on the same order of magnitude as the saturation intensity of the lasing transition.*

Let us revisit the threshold condition:

$$sc\sigma\Delta N_0 = \frac{\sigma_c}{\epsilon_0}$$

From Eq. 7.29 and noting that $cT_R = 2L$, we can rewrite the condition as

$$R_1 R_2 \exp\left[2Ls\sigma\Delta N_0 - 2\alpha l\right] = 1 \tag{7.38}$$

Finally, we examine graphically the steady-state solutions to the amplitude and phase equations, Eqs. 7.24 and 7.25, in the frequency domain:

$$\begin{aligned} \frac{\sigma_c}{2\epsilon_0} &= -s\frac{\omega}{2}\chi''(\omega) \\ (\omega - \Omega) &= -s\frac{\omega}{2}\chi'(\omega) \end{aligned}$$

In almost all cases, the gain width Γ is much broader than the spacing between two adjacent resonator resonances, as indicated by the vertical lines in Fig.7.3a, where we show ω_0 closer to the resonance on the left, which is the mode that will oscillate. The two equations are not independent, because χ' and χ'' are both dependent on the population inversion and ω. Close to ω_0, χ'' is nearly independent of ω whereas χ' is nearly linear in $\omega - \omega_0$. From Eq. 7.14, we have $\chi' = -[(\omega - \omega_0)/\Gamma]\chi''$. Since $\chi''(\omega) \approx \chi''(\omega_0)$, we have

$$s\chi'(\omega) \approx \frac{\sigma_c}{\omega_0\epsilon_0}\frac{\omega - \omega_0}{\Gamma}$$

and therefore

$$\omega - \Omega \approx (\frac{\sigma_c}{\omega_o\epsilon_0})\frac{\omega_0 - \omega}{\Gamma}$$

The left-hand side is the deviation of ω from the resonator frequency whereas the right-hand side is its deviation from the gain peak (Fig. 7.3b). The intersection of the two yields ω, which is in between Ω and ω_0. In practice, one often can tune Ω by maximizing the output laser power so that $\omega = \omega_0$.

Rate equations are powerful tools. We have considered only optical interactions, but with minor modifications, rate equations for other processes can be written down immediately. For example, in a gaseous medium where pumping is achieved by electron impact, the pumping rate can be written as σNF, where σ is the electron excitation cross section, N the atomic density, and F the electron flux, which is the electron density times electron velocity.

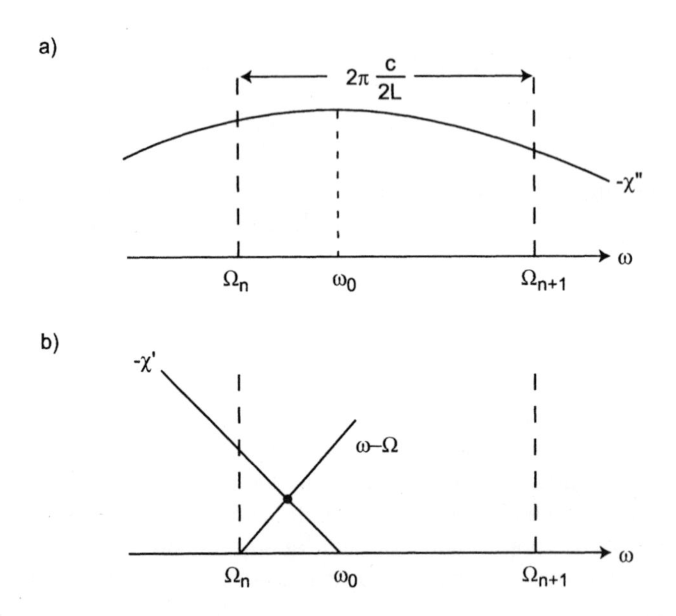

Figure 7.3: (a) Imaginary part of the susceptibility χ'' versus frequency. The gain is proportional to $-\chi''$. Two adjacent resonator modes are separated by 2π times the inverse round-trip time, or $2\pi\,[c/2L]$. (b) Real part of the susceptibility χ' versus frequency. The intersection of $-\chi'$ and $\omega - \Omega$ yields the lasing frequency.

7.4 Applications of the Rate Equations

We will now look at a few examples to illustrate how to apply the results obtained above to calculate the pumping thresholds and output powers of two lasers, and the gain of an amplifier. Most atomic media in lasers are too complex to be treated in full, and they must be simplified in order to keep focus on the essential processes. We have been discussing two-level atoms. However, it is impossible to have population inversion and therefore a laser in a medium with only two levels,[8] because stimulated emission would deexcite the atom to the lower level as soon as there was any inversion. A third level must be involved. The rate equations apply to the two lasing levels, as well as any other pair of levels interacting with light.

7.4.1 The Nd:YAG laser

The Nd:YAG laser is one of the most widely used lasers in engineering and scientific research. Pumped by a semiconductor laser, it is efficient ($\sim 10\%$). Its temporal and spatial output characteristics are close to ideal. It can be operated in a single-frequency mode for spectral purity, or mode-locked for a train of short pulses, or pulsed for a single, high-energy pulse, or frequency-doubled to pump other lasers. The active atoms are neodymium ions in the yttrium-aluminum-garnet (YAG) crystal, with a concentration $\simeq 1\%$. The energy levels are rather complicated, and only the levels that participate in the pumping and lasing action are drawn in Fig. 7.4a, with their spectroscopic notations. There are several possible pumping transitions, all optical, but we consider the pumping transition only at 0.8 μm, from the ground level 0 to level 3. From level 3, the ion quickly relaxes to level 2, the upper level of the lasing transition. The lower lasing transition, level 1, also has a very rapid decay time back to the ground state. The decay from level 2 to level 1 is almost completely by spontaneous emission with a much longer lifetime $T_1 \sim 10^{-3}$ s. The lasing transition line width Γ, caused by interaction between the ion and its surrounding vibrating atoms (phonons), is about $2\pi \times 1.3 \times 10^{11}$ s^{-1}. The transition wavelength is 1.06 μm. The stimulated emission cross section σ_E is calculated from Eq. 7.18 to be 10^{-18} cm^2.

We now calculate the threshold pump power requirement, and the output power of the laser. We leave the details of the resonator to Chapter 8, and simply assume that the field is confined to a cross-sectional area A inside the resonator. We will even ignore the refractive index of the YAG host. We assume that the main loss is the output mirror, which transmits 2% and reflects 98% of the lasing wavelength ($R_1 = 0.98$); the other mirror reflects 100% ($R_2 = 1$). The pumping can be longitudinal from one end, or transverse from the side. We will consider longitudinal pumping through one mirror that transmits 100% of

[8]The only exception is the excimer laser. The excimer molecule is formed in the upper state. The lower state is unstable from which the molecule dissociates into its component atoms.

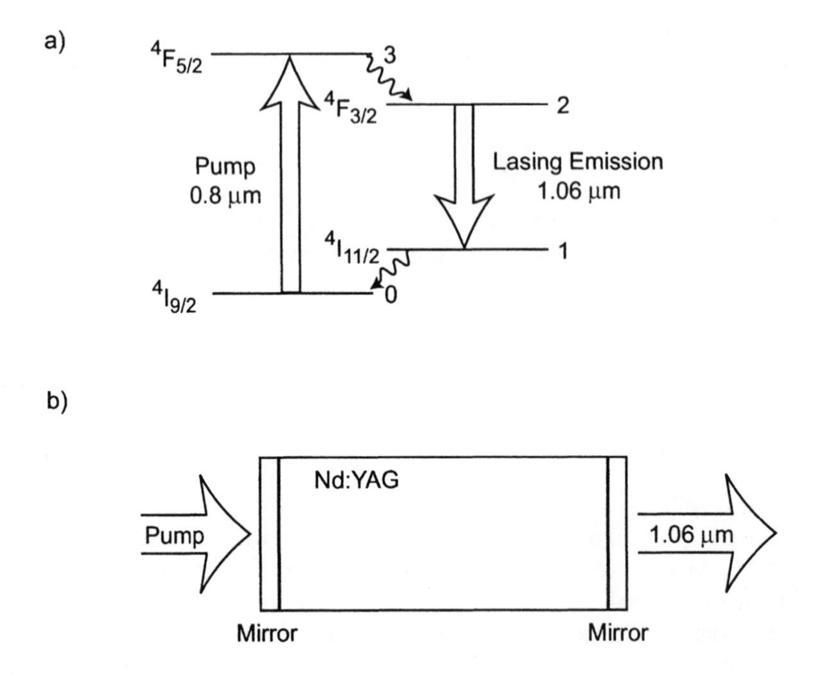

Figure 7.4: (a) Energy levels of Nd ions in YAG. The pumping is from ground level 0 to level 3, which relaxes quickly to level 2, the upper level of the lasing transition. The lower level of the lasing transition is level 1, which relaxes quickly to level 0. (b) Schematic of a longitudinally pumped Nd:YAG laser. The left mirror transmits the pumping wavelength but reflects the lasing wavelength. The right mirror transmits a small percentage of the lasing power.

the pump wavelength, with the pump beam cross-sectional area matching that of the lasing beam, and the YAG long enough to absorb all the pumping power. The simplified Nd: YAG laser is illustrated in Fig. 7.4b.

Let us first calculate the pumping power needed to reach lasing threshold. One could write down a rate equation for each of the four levels; however, because of the fast decays from levels 3 and 1, we can make the approximation that these two levels are empty. Ions excited to level 3 from the ground level by the pump light immediately decay to level 2, therefore the pump rate to level 2 is the rate at which the ions are pumped out of the ground level, or

$$\frac{\Delta N_0}{T_1} = \sigma_A \frac{I_{\text{pump}}}{\hbar\omega_A} N_0$$

where I_{pump} is the pump light intensity incident on the ion, σ_A is the cross section of the transition between levels 0 and 3, and $\hbar\omega_A$ is the photon energy of the pump. Because of the fast decay of the upper pump level 3, the pump transition is not saturated, and the pump beam decreases exponentially into the medium at the rate of $\sigma_A N_0$. We define the effective gain length to be the inverse of this rate, or $l = 1/(\sigma_A N_0)$. The gain per round trip at the lasing transition is $\exp(2\Delta N_0 \sigma_E l)$, which at threshold must be equal to $1/R_1 R_2 \simeq 1.02$. The factor $\Delta N_0 l$, from the equation above and the definition of l, is equal to $T_1 I_{\text{pump}}/(\hbar\omega_A)$, which yields the threshold pumping intensity of 2 W/cm^2. The pumping beam is focused onto the same area as the lasing beam area A; therefore the threshold pump power, denoted as $P_{\text{pump,th}}$, is $2A$ watts with A in cm^2.

To calculate the output power of the laser, we need the saturation intensity I_S of the lasing transition, $\hbar\omega_E/(\sigma_E T_1)$, where $\hbar\omega_E$ is the photon energy of the lasing transition at 1.06 μm. It is 200 W/cm^2. If the pumping power is P_{pump}, then the output power of the laser is, by Eq. 7.37

$$P_{\text{out}} = \frac{(1-R_1)}{2} A I_S \left(\frac{P_{\text{pump}}}{P_{\text{pump,th}}} - 1 \right)$$

For example, if we let A=0.2 cm^2, the threshold pumping power is 0.4 W. The output power, at twice the pumping threshold (0.8 W), is 0.3 W. The *differential* efficiency $dP_{\text{out}}/dP_{\text{pump}}$ can be easily calculated to be $\omega_E/\omega_A \simeq 0.8$, which means that one photon of the pump is converted into one photon of the lasing emission.

In the calculation just performed, the inverse absorption length of the pump $N_0\sigma_A$ is canceled out and is not needed. The value is needed for transverse pumping. At about 1% concentration, the absorption length is a few millimeters.

Problem 7.1 *The titanium:sapphire laser is one of the most widely used lasers in the laboratory. With a line width of* $2\pi \times 10^{14}$ s^{-1}, *it serves well both as a tunable single-frequency source and a femtosecond pulse source. The lasing* *transition, centered near 0.73 μm, is between two bands of vibration modes, as is the optical pumping transition. It can be modeled as a four-level system with an energy diagram similar to that of the Nd:YAG. Like the Nd:YAG, the lifetime from the upper pump level to the upper lasing level, and the lifetime of the lower lasing level, are much shorter than the lifetime of the upper lasing level (approximately equal to the radiative lifetime of 4 μs). Suppose the pump is from a frequency-doubled YAG laser (0.53 μm), and the resonator is designed so that the cross section of the laser beam in the Ti:sapphire material is 1 mm, and the output mirror transmits 5%. Find the threshold pumping power and the output power at 50% above threshold. Assume that the pumping transition cross section is approximately equal to the lasing transition cross section, and the titanium density is* 3×10^{19} *cm^{-3}.*

7.4.2 The erbium-doped fiber amplifier

Having seen a four-level system in the Nd:YAG laser and the Ti:sapphire laser, we now turn to a three-level system: the erbium-doped fiber *amplifier* (EDFA). An optical fiber is a cylindrical waveguide. The waveguide confines light to an inner core that has a slightly higher index of refraction. The EDFA is the key element in present fiber systems for long-distance high-data-rate transmission. By direct amplification of optical signals, it eliminates repeaters (which convert optical signals to electrical signals, then amplify them and convert them back to optical signals). The erbium ion, which provides optical gain, is a three-level system. This three-level system has one important difference compared to the four-level system, in that the ground state is the lower level of the amplifying transition. Since the atoms are normally in the ground state, and there cannot be gain unless the upper level is more populated than the lower, ground state, the three-level system cannot provide any gain until at least half of the atoms have been depopulated from the ground state. The gain threshold is therefore very high. Still, the erbium fiber amplifier is widely used because it provides gain at the 1.5 μm wavelength region where the loss is minimum in fibers. Erbium ions are embedded in the glass molecules of the fiber. The ground state and the first two excited states are shown in Fig. 7.5. Each state, labeled in the standard Russell–Saunders spectroscopic notation ${}^{2S+1}L_J$, actually consists of $2J+1$ sublevels, which are separated in energy through interaction with the electric fields of the surrounding glass molecules. An electron in one of these sublevels is scattered into other sublevels by interacting with the vibrations of the glass molecules (phonons). The net result of these two effects is that all the sublevels together appear like a single level whose energy width is approximately 4×10^{12} Hz. The main decay mechanisms of level 2 and 3 are different. Level 2 decays mainly by radiation to level 1, with a lifetime of about 10 ms. Level 3 decays mainly by phonon interaction to level 2, with a lifetime of about 20 μs, almost three orders of magnitude faster than the decay of level 2. From these parameters, the transition cross section between levels 2 and 1, σ_E, can be calculated to be 5×10^{-21} cm^2, and that between levels 1 and 3, σ_A, 2×10^{-21} cm^2. In application, a semiconductor laser emitting at 0.98 μm is coupled into

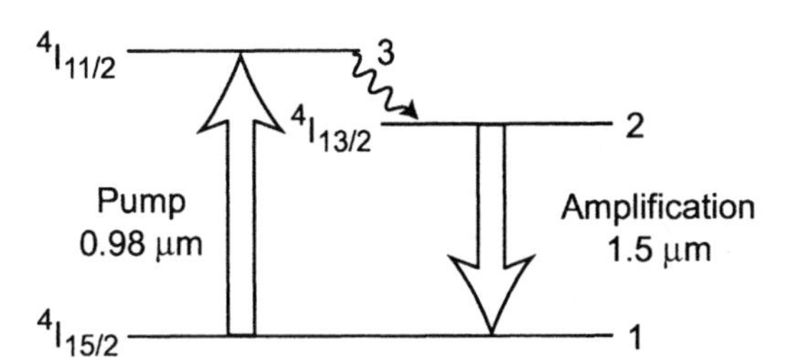

Figure 7.5: Energy levels of erbium ion in fiber. The lower level of the amplifying transition, level 1, is the ground level. Light of wavelength 0.98 μm pumps the ions from level 1 to level 3, from which they relax quickly to level 2, the upper level of the amplifying transition.

the fiber to pump the erbium ions from level 1 to level 3, from which the ions quickly decay to level 2, the upper level of the amplifier transition. Because ions in level 3 decay so quickly to level 2, we can make the approximation that the population of level 3 is zero. In the presence of a pumping beam of intensity I_P, and signal beam, which the amplifier amplifies, of intensity I_{sig}, the populations of levels 1 and 2 are changed by three mechanisms: decay from level 2 to level 1; stimulated emission and absorption of I_{sig}; and absorption of I_P. The rate equations for the population densities of levels 1 and 2, $N_{1,2}$ are then

$$\frac{dN_1}{dt} = \frac{N_2}{T_1} - \frac{\sigma_A I_P N_1}{\hbar\omega_P} - \frac{\sigma_E I_{\text{sig}}}{\hbar\omega_E}(N_1 - N_2) \simeq -\frac{dN_2}{dt}$$

where $\hbar\omega_P$ and $\hbar\omega_e$ are the energy difference between levels 3 and 1, and between levels 2 and 1, respectively. The total population density, N_0, remains constant, and is given by

$$N_0 = N_1 + N_2 + N_3 \simeq N_1 + N_2$$

From these two equations, we can solve for the populations in terms of light intensities, in the steady state:

$$N_1 = N_0 \frac{1 + I_{\text{sig}}/I_S}{1 + 2I_{\text{sig}}/I_S + I_P/I_{\text{th}}}$$

and

$$N_2 = N_0 \frac{I_P/I_{\text{th}} + I_{\text{sig}}/I_S}{1 + 2I_{\text{sig}}/I_S + I_P/I_{\text{th}}}$$

where we have introduced two normalization intensities, the saturation intensity I_S of the transition

$$I_S = \frac{\hbar\omega_E}{\sigma_E T_1}$$

and the threshold intensity I_{th}:

$$I_{\text{th}} = \frac{\hbar\omega_P}{\sigma_A T_1}.$$

The reason for the term *threshold intensity* will become clear. The small value of I_S (4 kW/cm^2) means that the amplifier is easily saturated. For a fiber whose radius is 5 μm, the saturation power is about 3 mW. The rate equations for the intensities are

$$\begin{aligned}\frac{dI_P}{dz} &= -\sigma_A N_1 I_P \\ \frac{dI_{\text{sig}}}{dz} &= \sigma_E (N_2 - N_1) I_{\text{sig}}.\end{aligned}$$

Substitution of N_1 and N_2 from above into these equations yields two coupled equations for the intensities:

$$\begin{aligned}\frac{dI_P}{dz} &= -\sigma_A N_0 \frac{1 + I_{\text{sig}}/I_S}{1 + 2I_{\text{sig}}/I_S + I_P/I_{\text{th}}} I_P \\ \frac{dI_{\text{sig}}}{dz} &= \sigma_E N_0 \frac{I_P/I_{\text{th}} - 1}{1 + 2I_{\text{sig}}/I_S + I_P/I_{\text{th}} I_{\text{sig}}}\end{aligned}$$

The second equation shows that, for the signal to grow, the right- hand side must be positive, (i.e., $I_P > I_{\text{th}}$). Because the gain threshold depends on pump *intensity*, EDFAs usually have smaller core areas than regular fibers to minimize the pump *power*. To find the gain of the amplifier, divide the first equation by the second:

$$\frac{dI_P}{dI_{\text{sig}}} = -\frac{\sigma_A}{\sigma_E} \cdot \frac{1 + I_{\text{sig}}/I_S}{I_P/I_{\text{th}} - 1} \cdot \frac{I_P}{I_{\text{sig}}}$$

Separating the two intensities and integrating from the initial to the final values (at $z = 0$ and $z = L$, respectively) of the intensities, we obtain

$$\frac{\sigma_E}{\sigma_A} \int_{I_P(0)}^{I_P(L)} dI_P \left(\frac{1}{I_P} - \frac{1}{I_{\text{th}}} \right) = -\int_{I_{\text{sig}}(0)}^{I_{\text{sig}}(L)} dI_{\text{sig}} \left(\frac{1}{I_{\text{sig}}} + \frac{1}{I_S} \right)$$

Although the integration is elementary, we make some simplifying assumptions. We assume that the length L of the fiber is chosen such that $I_P(L) \simeq I_{\text{th}}$. If the fiber is too much longer, then the signal will be absorbed again. If it is too much shorter, the amplifier will provide less gain than allowed by the input pump. We also assume that the initial pump intensity $I_P(0)$ is much greater than I_{th}. The above equation then yields

$$\frac{\sigma_E}{\sigma_A} \frac{I_P(0)}{I_{\text{th}}} \simeq \left[\ln \left(\frac{I_{\text{sig}}(L)}{I_{\text{sig}}(0)} \right) + \frac{I_{\text{sig}}(L) - I_{\text{sig}}(0)}{I_S} \right]$$

The gain can then be plotted for given input signal and pump intensities. The small-signal gain, when $I_{\text{sig}} \ll I_S$, can be cast in a simple form. In this case, for significant gain, the logarithmic term is much larger than the second term on the right-hand side of the equation above, and

$$I_{\text{sig}}(L) \simeq I_{\text{sig}}(0) \exp\left(\frac{\sigma_E}{\sigma_A} \frac{I_P(0)}{I_{\text{th}}}\right)$$

The gain for signal is often expressed in decibels per watt of pump power:

$$10 \log_{10}\left[\frac{I_{\text{sig}}(L)}{I_{\text{sig}}(0)}\right] \simeq 10 \left(\frac{\sigma_E}{\sigma_A} \frac{I_P(0)}{I_{\text{th}}}\right) \log_{10}(e)$$

For an EDFA with a core radius of 5 μm, the gain is about 1 dB/0.8 mW; an amplifier providing 30 dB of small-signal gain requires a pump power of 24 mW. Now, to find the length of the fiber, we integrate the differential equation for I_P in the absence of the signal, the length L of the EDFA is given approximately by $\sigma_A N_T L \simeq I_P(0)/I_{\text{th}}$. For a doping density N_T of 10^{18} cm^{-3}, an input pump power of 24 mW, and a core radius of 5 μm, L is about 15 m.

Problem 7.2 *A piece of erbium-doped fiber of length L is joined at the ends to form a loop. The fiber is pumped by a laser at* 0.98 μm *to make a fiber laser. The pump is coupled through another piece of undoped fiber that is placed closely to the loop. The coupler transmits 100% of the pump and 10% of the erbium lasing wavelength of* 1.5 μm. *An isolator is placed within the loop so that light can travel in only one direction. At* 1.5 μm, *the attenuation in the fiber due to Rayleigh scattering is* 0.2 *dB/km. If L is approximately the absorption length of the pump, what is the threshold pump power? What is the output power at twice the pumping threshold? Use the data in the EDFA example.*

7.4.3 The semiconductor laser

The semiconductor laser is probably the most important commercial laser. It is used in fiber communication systems, as well as many other applications. It is also the smallest laser, only a fraction of a millimeter long. It comes in many different structures, and with different materials, operates at different wavelengths. The most important wavelength bands are 0.8, 1.3, and 1.5 μm. It is an optoelectronic device with a diode pumped by the current passing through it. The optical transition is between two bands of levels: the upper band, *conduction* and the lower, *valence.* Gain is provided by electrons going from the bottom of the conduction band to the top of the valence band. The theory developed so far for two-level atoms is, strictly speaking, not applicable; at the least, it must be extended to include the distributions of levels in the two bands. Furthermore, the interaction between light and matter has been assumed to be dipolar, which means that the wavefunctions of the matter do not extend to a significant part of the optical wavelength. This is not true in a semiconductor, where the electronic wavefunctions are extended. Nevertheless, it is found that

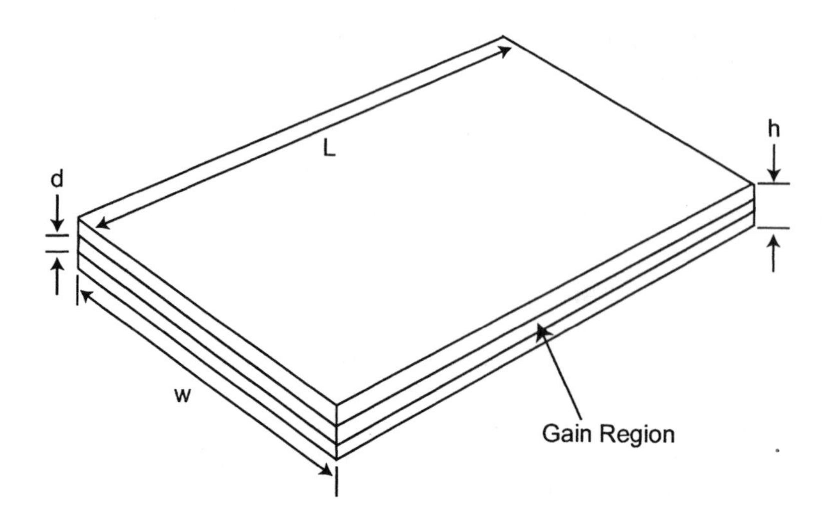

Figure 7.6: Schematic of a semiconductor laser. Electrons in a current passing through the laser from top to bottom make transitions in the gain region from the conduction to the valence band.

the gain per unit length in a semiconductor is approximately proportional to the electron density N in the conduction band. One can then define a transition cross section σ as the proportionality constant so that the gain per unit length is σN. σ depends on frequency, material, and the structure of the diode, but is typically in the neighborhood of 10^{-16} cm^2. Electrons in the conduction band decay back to the valence band, mostly by spontaneous emission, with a typical lifetime T_1 of a few nanoseconds. Because of the great variety of devices available and the complexity of these devices, a fair treatment of this laser requires a specialized book. Here, we will be contented with obtaining order-of-magnitude values for some important parameters under typical conditions.

We consider the structure illustrated in Fig. 7.6 It is a typical quantum-well laser. The optical resonator consists of a dielectric waveguide within which most of the light is confined. Its width is w, typically a few micrometers; its height is h, typically 1 μm or less; light travels along a length of L, typically a few hundred micrometers. At the ends of the waveguide, light is reflected by the semiconductor/air interface, with a reflectivity R of about 30%, although in some devices higher reflectivities are obtained by coating. The gain is provided by a quantum-well layer inside the waveguide. The thickness d of the quantum-well layer is typically about 10 nm. Surrounding the waveguide are semiconductor materials to make up the rest of the diode and guiding structure, with which we will not be concerned. An electric current passing through the waveguide excites the electrons in the quantum well from the valence band to the conduction band and provides the optical gain. We will first calculate the threshold current i_{th}. The threshold condition is given by Eq. 7.38, in which we insert the filling factor d/h and ignore the loss α

$$R^2 \exp\left(2\frac{d}{h}\sigma N_{\text{th}} L\right) = 1$$

or

$$2\frac{d}{h}\sigma N_{\text{th}} L = \ln\left(\frac{1}{R^2}\right) \tag{7.39}$$

To relate the threshold electron density N_{th} to the current, we note the current is simply the change of charge in time. The total number of electrons in the quantum well, each with charge $-e$, is $N_{\text{th}} dwL$. The electrons decay with a time constant T_1. So the threshold current is

$$\begin{aligned} i_{\text{th}} &= \frac{eN_{\text{th}} dwL}{T_1} \\ &= \frac{e}{T_1}\frac{wh}{\sigma}\ln\left(\frac{1}{R}\right) \end{aligned}$$

We next calculate the output power at a pumping current i. Since i is proportional to ΔN_0, from Eq. 7.36, the intensity inside the resonator is

$$I = I_S\left(\frac{i}{i_{\text{th}}} - 1\right)$$

where I_S is the saturation power $\hbar\omega/(\sigma T_1)$. The power inside the resonator is I times the cross-sectional area of the waveguide wh. This power multiplied by the mirror transmission $(1 - R)$ is the output power:

$$P_{\text{out}} = (1 - R) wh I_S \left(\frac{i}{i_{\text{th}}} - 1\right)$$

For $w = 3$ μm, $h = 1$ μm, $L = 200$ μm, $R = 30\%$, $\sigma = 5 \times 10^{-16}$ cm^2, $T_1 = 3$ ns, and at a wavelength of 1.5 μm, $i_{\text{th}} = 3.6$ mA, and at twice the threshold current, P_{out} is about 2 mW. One common measure of the performance of the laser is the change of output power versus the change of current:

$$\frac{dP_{\text{out}}}{di} = (1 - R)\frac{whI_S}{i_{\text{th}}} = \left(\frac{1 - R}{\ln(1/R)}\right)\left(\frac{\hbar\omega}{e}\right)$$

The second factor $\hbar\omega/e$ on the right-hand side means one photon per electron, which is the most that can be obtained in this device. For $R = 30\%$, the laser output increases by 0.48 mW when the current is increased by 1 mA. Some semiconductor lasers have their facets coated to increase the mirror reflectivities. The first factor on the right-hand side approaches unity when R approaches unity.

Problem 7.3 *Suppose that the pumping term in the rate equation for population inversion is*

$$\Delta N_0 = \Delta N_{DC} + \delta N_0 \cos(2\pi f t)$$

where the first term on the right is time-independent and much larger than the second term. Apply perturbation methods to the rate equations and find the frequency response of the light intensity. Use the data in the section on the semiconductor laser, Section 7.4.3, to calculate the frequency at which the response is down 3 dB from the zero-frequency value. Semiconductor lasers are used to transmit data up to 10 gigabits per second by modulating their pumping currents. Is your answer consistent with this fact?

7.5 Multimode Operation

So far, we have discussed only *homogeneously broadened* laser media, whose atoms all have the same resonance frequency and interact with the same electromagnetic field. We saw that, by gain saturation, the gain at frequencies away from the oscillating frequency is less than the loss, and therefore fields at those frequencies are prevented from oscillation. Thus the laser oscillates in one mode. There are several circumstances under which a laser can oscillate in more than one mode. Depending on the application, multimode operation can be desirable or even necessary, as in the *mode-locked* laser to generate ultrashort pulses; or undesirable, as in optical fiber communication; or unimportant, as in a laser pointer. Even in a homogeneously broadened laser, multimode operation can occur, because the fields of different modes have different spatial distributions and therefore interact with atoms at different locations. Spatially selective saturation of the atomic gain due to the field distribution is called *spatial holeburning*.

7.5.1 Inhomogeneous broadening

Inhomogeneously broadened laser media have atoms with different resonance frequencies, the origin of which can be the Doppler effect as in a gaseous laser like the helium–neon, or different environments the atoms are in as in the solid state Nd:glass laser. Because of the different resonance frequencies, different laser modes, with different frequencies, interact with different groups of atoms and multimode operation is a natural outcome. Consider, for example, the helium–neon laser. The lasing that produces the familiar red beam at 0.63 μm is between the $5s$ and $3p$ levels of the neon atoms in the gas. The density of the gas is low enough so that collisions do not contribute significantly to the line width, and the dominant broadening mechanism of the transition is Doppler broadening. As discussed in Chapter 4, Section 4.6.3, in a gas at temperature T, the atoms move randomly with a kinetic energy roughly up to $k_B T$, where k_B is the Boltzmann constant. When an atom moves with a velocity v in the axial direction of the resonator, the transition resonance frequency is shifted, to an observer stationary with the tube, by an amount $(v/c)f_0$ where f_0 is the resonance frequency of the atom at rest. The spread of the velocity Δv is given by $k_B T = (\Delta v)^2/(2M)$, where M is the mass of the atom; and the spread of the resonance frequency Δf is equal to $(\Delta v/c)f_0$. For the HeNe laser, Δf ~2 GHz. The axial mode frequencies of the laser resonator are separated by $c/2L$, where L is the length of the resonator.[9] Now if $\Delta f > c/2L$, and $c/2L$ is in turn greater than Γ, the homogeneous line width of the transition, then each mode interacts with a group of atoms whose Doppler-shifted resonance frequency coincides with the mode frequency, within one homogeneous line width. For the HeNe laser, Γ is the spontaneous emission rate, approximately 100 MHz, and L is typically about 30 cm, so that $c/2L$ ~500 MHz. The distribution of the atoms as a function of resonance frequency is depicted in Fig. 7.7a. The unsaturated gain follows the atomic inversion and is shown in Fig. 7.7b. For a mode to be excited, the unsaturated gain must exceed the loss, so the oscillating modes are confined within the frequency range between the points where the unsaturated gain intercepts the loss (Fig. 7.7b). At every interval of $c/2L$ within that range, there is an oscillating mode and the saturated gain is equal to the loss (Fig. 7.7b). The output spectrum is shown in Fig. 7.7c. Between two modes, there is no optical field and the gain is not saturated and hence retains the unsaturated value and the saturated gain dips to the loss level at resonator resonances. This phenomenon is known as *spectral hole - burning*. The different modes are normally independent of each other, as they interact with different groups of atoms. The phase of each mode fluctuates slowly and randomly, which yields the line width δf of that mode (see Appendix 7.B). The relative phase between the modes varies slowly in time; after a time of $\sim 1/\delta f$, it will have changed completely. The total output is the superposition of these modes. It is not constant in time but varies periodically with a period of $2L/c$. The shape of the waveform changes in a time $\sim 1/\delta f$, as the relative phase changes completely in that time.

7.5.2 The mode-locked laser

To generate short pulses in time, by Fourier transformation, many modes at different frequencies are required. Moreover, to obtain the shortest pulses reproducibly, these modes must be locked in phase. It is interesting that, to date, the shortest pulses are obtained, by the method of *mode locking*, from homogeneously broadened lasers, whose natural tendency is single-mode operation, rather than from inhomogeneously broadened lasers which tend to emit many frequencies. The reason behind this seeming paradox is that in the homogeneously broadened laser, the modes are generated by a coherent process and are automatically excited in phase, whereas in an inhomogeneously broadened laser, the modes are present, without the definite phase relationship with one another that the mode-locking process has to create and maintain.

[9] The theory of resonators is discussed in detail in Chapter 8, but the mode separation frequency can be easily derived by accepting the fact that, because of the boundary conditions at the end mirrors, there must be an integral number of half wavelengths within L, each number corresponding to a mode.

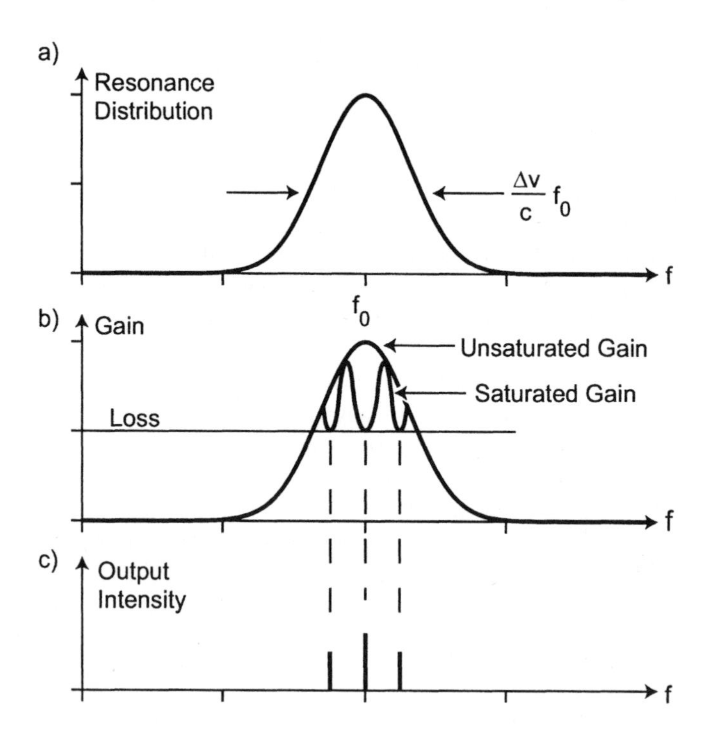

Figure 7.7: (a) Distribution of moving atoms versus resonance frequency. The peak frequency f_0 is the resonance frequency of atoms at rest. (b) Unsaturated and saturated gain versus frequency. Atoms whose resonance frequency coincides with a resonator mode lase in that mode, and the gain at that frequency saturates to the loss level. Three lasing modes are shown. (c) The light output spectrum for the gain in (b).

Quantitative theories of mode locking are beyond the scope of this chapter. Only a qualitative description is given here. There are many ways to excite and phase lock the axial modes of a laser. The most straightforward method is to place a modulator inside the resonator, which modulates the loss of the resonator at the round- trip frequency of the resonator (Fig. 7.8). As light passes through the resonator, the part that sees the least loss in one round trip will see the least loss in subsequent trips and therefore will be amplified most. Similarly, the part that sees the most loss will be attenuated in each subsequent round trip, and a pulse is formed that circulates inside the resonator. The pulse cannot be narrowed indefinitely as there are elements inside the resonator that broaden the pulse, such as the finite gain bandwidth or dispersive optical components. When the modulation is induced by the light pulse itself such that higher intensity suffers less loss, we have a very efficient pulse narrowing process called *passive mode locking*. One such mechanism that can be used to this effect is the nonlinear refractive index of a medium. The refractive index changes with light intensity. By itself the nonlinear refractive index is not lossy. However, in passing through the nonlinear medium, the peak of a pulse sees a refractive index different from that of the wings, and therefore the divergence angles are different at the peak and at the wings. If the nonlinearity is chosen properly so that more intense light diverges less than light at lower intensity, then, through an aperture, the lower intensities at the wings will be filtered out, and the pulse will be sharpened (Fig. 7.9).

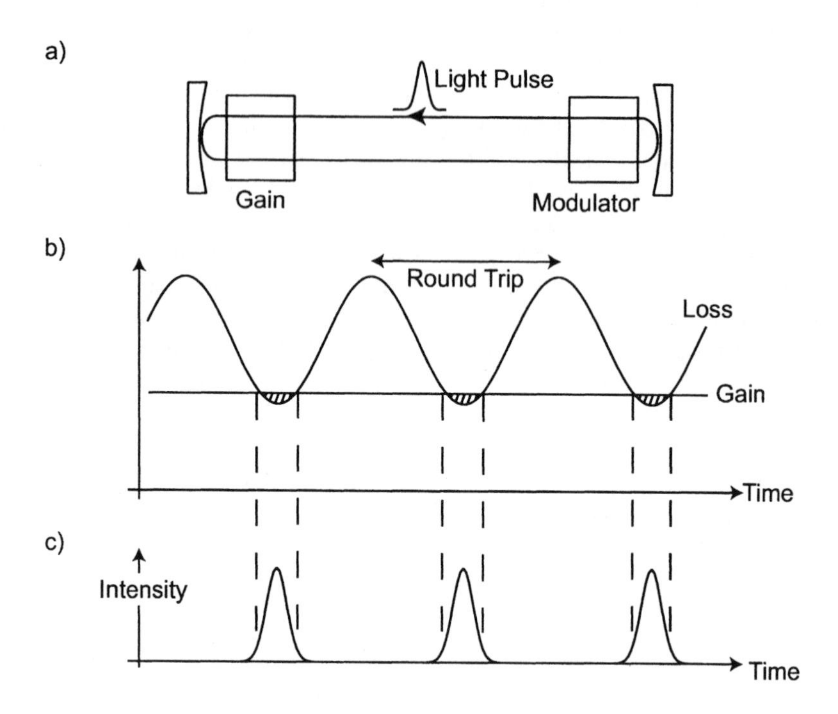

Figure 7.8: Mode locking of a laser to generate short pulses. Upper diagram: a light pulse circulates inside the resonator, which contains a gain medium and a modulator. The modulator is a device that attenuates light periodically in time, with a period equal to the round-trip time in the resonator. Middle trace: time variation of the modulator loss and saturated gain versus time. The gain is assumed to be saturated by the average intensity, and the period of the loss is equal to the round-trip time in the resonator. Near the minima of loss, the saturated gain exceeds the loss, and in between the minima the loss exceeds the gain. Light passing through the modulator near the loss minimum will be amplified repeatedly; light passing through the modulator between loss minima will be attenuated repeatedly. Lower trace: the pulsetrain generated.

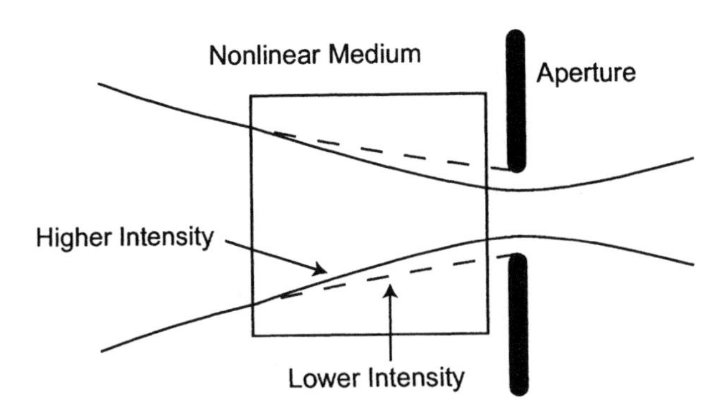

Figure 7.9: An aperture place after a nonlinear medium. The index of refraction increases slightly with light intensity so that the peak of a pulse diverges less than the wings. The aperture transmits more of the peak than the wings.

7.6 Further Reading

The laser theory developed in this chapter follows the line in

- M. Sargent III, Marlan O. Scully, and Willis E. Lamb, Jr., *Laser Physics*, Addison-Wesley, Reading, MA, 1974.

A thorough treatment of the similarity between lasers and classical oscillators can be found in

- A. E. Siegman, *Lasers*, University Science Books, Mill Valley, CA, 1986.

An excellent book that describes many laser systems, with an elementary yet careful and thorough treatment of the semiconductor laser is

- O. Svelto, *Principles of Lasers*, 4th edition, Plenum Press, New York, 1998.

Laser mode - locking and injection - locking are treated in

- H. A. Haus, *Waves and Fields in Optoelectronics*, Prentice-Hall, Upper Saddle River, NJ, 1984.

Extensive references on ultra-short-pulse lasers and techniques can be found in

- L. Yan, P.-T. Ho, and C. H. Lee, *Ultrashort Laser Pulses*, in *Electro-optics Handbook*, 2nd edition, R. Waynant and M. Edinger, eds., Academic Press, Boston, 2000.

A thorough treatment on ultra-short laser pulses and techniques is

- J.-C. Diels and W. Rudolph, *Ultrashort Laser Pulse Phenomenon: Fundamentals, Techniques and Applications on a Femtosecond Time Scale*, Academic Press, Boston, 1996

The theory of electronic oscillators follows

- K. Kurokawa, *An Introduction to the Theory of Microwave Circuits*, Academic Press, Boston, 1969.

Another treatment of the theory of the electron oscillator and its frequency width can be found in the following treatise, which also has a very insightful discourse on spontaneous emission, and a delightful quantum-mechanical theory of vacuum electronic oscillators:

- A. B. Pippard, *The Physics of Vibration*, Cambridge University Press, Cambridge, UK, 1989,

Appendixes to Chapter 7

7.A The Harmonic Oscillator and Cross Section

7.A.1 The classical harmonic oscillator

In the semiclassical theory of light–matter interaction, quantum mechanics comes in only in the treatment of matter, here simplified to a two-level system. Transition between the two levels can be modeled as a quantum harmonic oscillator. The key quantity in the quantum mechanical theory of light–matter interaction is the complex susceptibility χ. It is interesting to see how to obtain the same result, heuristically, from the classical harmonic oscillator. The rate equations for the population density can be obtained from energy conservation.

Consider a particle of mass m and charge q sitting on a spring under an electric field $E(t) = \frac{1}{2}(E_0 e^{-i\omega t} + E_0 e^{i\omega t})$. The equation of motion for the displacement x of the particle from equilibrium is, by Newton's law

$$m\frac{d^2x}{dt^2} + m\omega_0^2 x = qE(t)$$

where ω_0 is the natural resonance frequency of the mass–spring system. The macroscopic polarization $P = \frac{1}{2}(P_0 e^{-i\omega t} + P_0^* e^{i\omega t})$ of a collection of these charges is the statistical average of Nqx, where N is the number of oscillators per unit volume. Multiplying the equation above by Nq and statistically averaging over the oscillators, we obtain the equation for the macroscopic polarization

$$\frac{d^2P}{dt^2} + 2\Gamma\frac{dP}{dt} + \omega_0^2 P = N\frac{q^2}{m}E(t)$$

where we have introduced a dephasing time constant Γ. We now cast the innocent-looking factor q^2/m on the right-hand side into a different form. In classical physics, whenever a charge is accelerated, it radiates. The time-averaged power P_{rad} radiated by an oscillating dipole is, from classical electrodynamics,

$$P_{\text{rad}} = \frac{q^2x^2\omega^4}{12\pi\epsilon_0 c^3}$$

which is a special case of Larmor's formula.[10] The energy stored in a harmonic oscillator is $m\omega^2x^2/2$. The classical radiative rate γ_e can be defined as the ratio $P_{\text{rad}}/(m\omega^2x^2/2)$ or

$$\gamma_e = \frac{q^2\omega^2}{6\pi\epsilon_0 mc^3}$$

In terms of γ_e, the factor q^2/m is equal to $(6\pi\epsilon_0 c^3/\omega^2)\gamma_e$, and the equation for P becomes

$$\frac{d^2P}{dt^2} + 2\Gamma\frac{dP}{dt} + \omega_0^2 P = N\frac{6\pi\epsilon_0 c^3}{\omega^2}\gamma_e E(t)$$

[10] We have already seen this radiated power expression in terms of the transition dipole matrix element μ_{12} in Eq. 4.49 at the end of Section 4.4.2. Note that the matrix element $\mu_{12} = \frac{1}{2}|\mu| = \frac{1}{2}|qx|$, the oscillating dipole itself.

This equation yields the classical susceptibility, which has exactly the same form as that of the quantum-mechanical oscillator if γ_e is replaced by the spontaneous emission rate $\gamma = A_{21}$. This emission rate γ can be obtained in a heuristic way by using Larmor's formula once again. The energy stored in a quantum-mechanical oscillator is $\hbar\omega$. The power lost by spontaneous emission is $\hbar\omega\gamma$. If we equate $\hbar\omega\gamma$ to P_{rad}, identifying qx with the dipole moment $2\mu_{12}$, we obtain

$$\gamma_e \to \gamma = \frac{\mu_{12}^2\omega^3}{3\pi\epsilon_0 c^3\hbar}$$

The classical harmonic oscillator always absorbs energy from the field. We know that if we have population inversion, the reverse happens. We therefore replace oscillator density N by $N_1 - N_2 \equiv -\Delta N$, with which the right-hand side of the equation for P becomes

$$-\frac{2\omega}{\hbar}\mu_{12}^2\Delta N E(t)$$

and a susceptibility is immediately obtained which is the same obtained in Chapters 2 and 7.

7.A.2 Cross section

The concept of interaction cross section is very useful in visualizing, and convenient in quantifying, the strength of an interaction. The cross section is an imaginary surface area, although in rare cases it may be the same as some physical surface area, like the dish antenna used to receive satellite TV signals. The stimulated emission cross section was derived in Chapters 2 and 7. Here, as a simple example, we derive the cross section of scattering of a plane electromagnetic wave by a single, free electron.

Consider an otherwise free electron in the electric field of an incident plane wave. The equation of motion of the electron is:

$$m\ddot{x} = -eE_0 e^{-i\omega t}$$

The power radiated by the electron, averaged over one cycle, is given by Larmor's formula above:

$$P_{\text{rad}} = \frac{q^2\ddot{x}^2}{12\pi\epsilon_0 c^3}$$

The incident wave intensity (power/area) I, averaged over one cycle, is

$$I = \frac{1}{2}\epsilon_0 c E_0^2$$

If we divide the scattered power by the incident intensity, we get a measure of the strength of the scattering process. The ratio, which has the dimension of area, is the scattering cross section for the free electron, σ_{fr} :

$$\sigma_{\text{fr}} = \frac{8\pi}{3}\left(\frac{e^2}{4\pi\epsilon_0 mc^2}\right)^2 \equiv \frac{8\pi}{3}r_e^2$$

We have defined the quantity within the brackets to be r_e, which is the classical electron radius. If we imagine that the charge of the electron is distributed uniformly over a spherical surface of radius r_e, then the stored potential energy is, to within a factor of two, $e^2/(4\pi\epsilon_0 r_e)$. Equating this energy to the energy of the rest mass of the electron mc^2 yields the formula for r_e, which can be evaluated to be about 3×10^{-15} m. The cross section is a fictitious surface presented to the incident plane wave by the electron. The power incident on that surface is absorbed and reradiated, that is, scattered by the electron. To illustrate the application of this physical interpretation, we calculate the attenuation of the incident wave intensity by scattering. Suppose that there are N electrons per unit volume. Along the wave propagation direction z, the total number of electrons contained in an infinitesimal volume of area A and width dz is $NAdz$. Each electron presents an area σ_{fr} to the wave, so that the total fictitious, absorbing surface presented to the incident wave is $\sigma_{\text{fr}} NAdz$, and the energy absorbed from the incident wave is $I(z)\sigma_{\text{fr}} NAdz$. The incident power is $I(z)A$, and the power, after passing through the volume, is $I(z+dz)A$. By energy conservation, the difference between them must be equal to the power absorbed by the electrons

$$I(z)A - I(z+dz)A = I(z)\sigma_{\text{fr}} NAdz$$

or

$$\frac{dI}{dz} = -\sigma_{\text{fr}} NI(z)$$

The attenuation coefficient per unit length is $\sigma_{\text{fr}} N$.

Now, if we add a restoring force and a damping term to the electron equation of motion and go through the process again, we will find that the scattering cross section σ_{ho} of the harmonic oscillator is enhanced over that of the free charge:

$$\sigma_{\text{ho}}(\omega) = \sigma_{\text{fr}} \left[\frac{\omega^2/4}{(\omega - \omega_0)^2 + (\gamma/2)^2} \right]$$

At resonance $\omega = \omega_0$, we obtain

$$\sigma_{\text{ho}}(\omega_0) = \sigma_{\text{fr}} Q^2$$

where $Q \equiv \omega_0/\gamma$ is the quality factor of the oscillator, which is the number of cycles the oscillator undergoes before $1/e$ of its energy is dissipated. The apparent radius of the harmonic oscillator is increased by a factor of Q over that of the free electron. Now if the damping is caused by scattering alone, $\gamma = \gamma_e$, then

$$\sigma_{\text{ho}}(\omega_0) = \frac{3}{2\pi}\lambda^2$$

which is the same as Eq. 7.19.

Problem 7.4 *(a) Express the classical radiative lifetime γ_e in terms of the transit time through a classical electron radius and the oscillation frequency. What is γ_e at 0.5 μm wavelength?*

(b) Calculate the spontaneous emission lifetime at 0.5 μm when μ is (i) a_0; (ii) $0.01a_0$, where a_0 is the Bohr radius.

Problem 7.5 *In Chapter 6, the dissipative force on the atom is interpreted as proportional to the rate of absorption of photon momentum, $\hbar\mathbf{k}$. The force can also be interpreted classically as discussed in Chapter 4, section 4.4.2. In classical electrodynamics, the Poynting vector divided by the speed of light is the radiation pressure $\mathbf{P}$, which has the dimensions of force per unit area. Show that under a plane wave, the force on the atom, Eq. 4.48, can be cast in the form*

$$\mathbf{F}_{abs} = \mathbf{P}\sigma$$

where σ, the effective area, is equal to the interaction cross section.

7.B Circuit Theory of Oscillators and the Fundamental Line Width of a Laser

7.B.1 The oscillator circuit

At a fundamental level, the laser does not differ from an electronic oscillator in its function of generating a coherent signal, except that, because of its short wavelength relative to the device size, the output of the laser is spatially confined and its spatial properties must be considered. Here we develop a circuit theory of oscillators using a negative resistance as the gain. *Negative resistance* in an electronic element refers to a negative *differential* of voltage versus current at some bias voltage or current. At the bias point of negative resistance, energy is transferred from the element to the rest of the circuit. It is intuitively acceptable, as a positive resistance leads to energy dissipation. The biasing circuit, unnecessary for the following discussion, is omitted from the electronic oscillator circuit.

The oscillator circuit model, shown in Fig. 7.10, consists of a resonator (the inductor L and the capacitor C), a gain element (the negative resistance $-R_g - iX$), a positive resistance R which represents the output coupling from the resonator, and an injection voltage source v_s. We have added a reactive (imaginary) part X to the negative resistance, which plays the same role as the real part of the susceptibility of an atomic transition. When X is nonzero, the oscillating frequency will deviate from the resonance frequency of the $L-C$ resonator. The real part of the negative resistance, R_g, depends on the amplitude of the current passing through it, as energy conservation requires it to be saturable. Both R_g and X are frequency-dependent. The injection signal v_s is used to represent two sources: (1) an outside signal used to lock the oscillator frequency, a process called injection locking and (2) noise. In either case, the injection signal is treated as a perturbation, a method familiar to students of quantum mechanics.

To find the current flowing in the circuit loop, we assume, as in light–matter interactions, an almost purely harmonic signal. The current I and the voltage

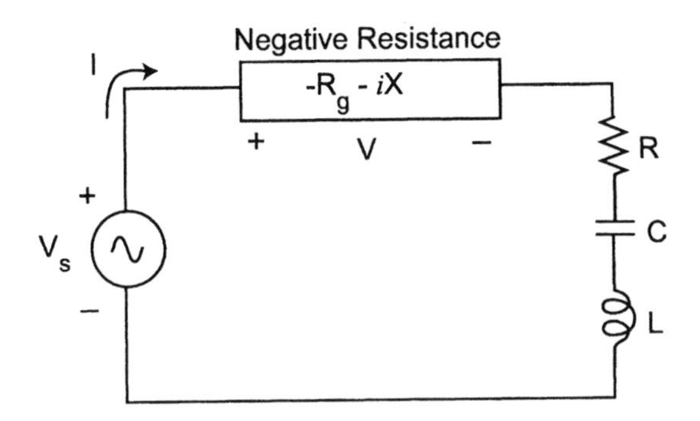

Figure 7.10: Circuit diagram of an electronic oscillator. The voltage source can be an injection signal or noise. Gain is represented by the negative resistance. Resonance is provided by the inductor and the capacitor. The resistor represents coupling loss in the oscillator.

across the negative resistance, V, are defined in Fig. 7.10. From elementary circuit theory, the equations for I is

$$L\frac{d^2I}{dt^2} + R\frac{dI}{dt} + \frac{I}{C} = -\frac{dV}{dt} + \frac{dv_s}{dt}$$

We assume

$$V = (-R_g - iX)I$$

which means that the negative resistance reacts fast enough to follow the current. The current I is to take the form

$$I(t) = A(t)\exp[-i\omega t - i\phi(t)]$$

where ω is the oscillating frequency, and the amplitude A and phase ϕ are real, varying slowly in one period. Ignoring second time derivatives of A and ϕ, and separating the real and imaginary parts, we obtain from the equation for I two equations for A and ϕ

$$\frac{dA}{dt} + \frac{R}{2L}A = \frac{R_g}{2L}A + \frac{1}{2L}\mathrm{Re}\left\{v_s(t)\exp[i\omega t + i\phi]\right\} \quad (7.40)$$

$$\frac{d\phi}{dt} + (\omega - \Omega) = \frac{X}{2\omega L} + \frac{1}{2LA}\mathrm{Im}\left\{v_s(t)\exp[i\omega t + i\phi]\right\} \quad (7.41)$$

where $\Omega = \sqrt{1/(LC)}$ is the resonance frequency of the $L - C$ resonator. Note the similarity of these equations to Eqs. 7.24 and 7.25 and the corresponding physical quantities. We now solve the equations in three particular cases.

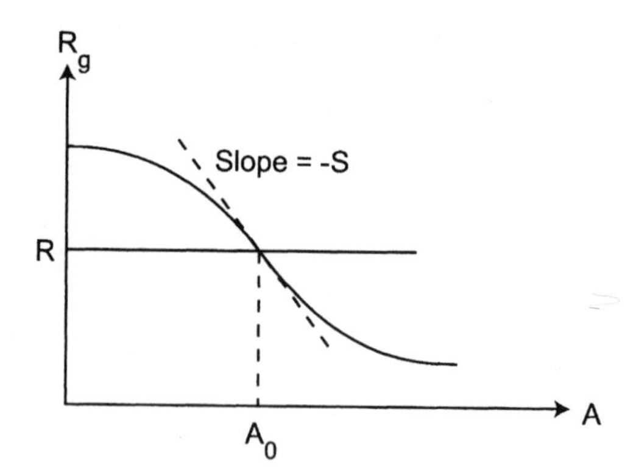

Figure 7.11: Gain (negative resistance R_g) versus amplitude A of current passing through it. The steady-state A_0 is obtained at the point where the saturated gain equals the loss R. The slope at that point is negative, so that slight fluctuations in A are damped.

7.B.2 Free-running, steady-state

In this case, there is no injection signal, $v_s = 0$. The equations above then become

$$\begin{aligned} R &= R_g(A) \\ (\omega - \Omega) &= \frac{X}{2\omega L} \end{aligned}$$

The first equation says that the positive resistance is equal to the negative resistance, that is, loss is equal to gain. The second equation determines the steady-state oscillation frequency ω. In general, the negative resistance decreases with increasing current amplitude A, as shown in Fig. 7.11, the intersection of R with $R_g(A)$ determines the steady-state oscillation amplitude A_0.

7.B.3 Small harmonic injection signal, steady-state

A small, pure harmonic injection signal

$$v_s(t) = v_0 \exp(-i\omega_s t)$$

is applied with frequency ω_s, which may be different from the free-running oscillation frequency ω_0 determined above. The question is whether the oscillator can be locked to the external injection signal and oscillate at the injection frequency. This is a common and useful technique in locking several oscillators to a reference. For example, the oscillator ("slave") may not be as stable, or as pure, as the injection ("master"). It is possible, by this method, to obtain a better, more powerful source from a weaker, more stable source, in electronics as well as in lasers.

We assume that injection is successful, and the oscillator is in the steady state, oscillating at the injection frequency ω_s. The amplitude A and frequency ω_s of the current differ slightly from those free running

$$A = A_0 + \Delta A,\ \omega_s = \omega_0 + \Delta\omega$$

where the subscript 0 denotes free-running values and Δ denotes deviations caused by injection. In applying perturbation theory to Eqs. 7.40 and 7.41 we expand the negative resistance at A_0 :

$$R_g(A_0 + \Delta A) \simeq R_g(A_0) + \frac{dR_g}{dA}\Delta A \equiv R_g - s\Delta A.$$

The derivative $-s$ is negative, indicating saturation, as shown in Fig. 7.11. For simplicity, we ignore the change of X with respect to A and ω. With this definition, we have from perturbation theory,

$$\begin{aligned} \Delta\omega &= \frac{v_0}{2LA_0}\sin(\phi) \\ sR\Delta A &= \frac{v_0}{2L}\cos(\phi) \end{aligned}$$

where ϕ is the phase difference between the free-running current and the injection signal. These two equations are plotted in Fig. 7.12. Since $|\sin\phi| \leq 1$, the maximum locking range of $\Delta\omega$ is

$$|\Delta\omega|_{\max} = \left|\frac{v_o}{2LA_0}\right|$$

We can write this equation in terms of more general physical quantities by multiplying and dividing the right-hand side by the output coupling resistance R. The term $A_0 R$ is the output voltage, and $R/L \equiv \Delta\omega_e$ is the frequency width of the "cold" resonator formed by the passive elements R, L, and C. Since a voltage is proportional to the square root of power P, we have

$$|\Delta\omega|_{\max} = \frac{\Delta\omega_e}{2}\sqrt{\frac{P_{\text{injection}}}{P_{\text{output}}}}$$

This equation for the maximum locking range is called the Adler equation. Since by assumption $P_{\text{injection}} < P_{\text{output}}$, the frequency locking range is smaller than the cold resonator width.

The quantity ΔA is positive within the locking range ($|\phi| \leq \pi/2$), which means that the power under injection is higher than that free-running, and that the saturated gain R_g is less than that free-running. Under these conditions the saturated gain is less than the loss, the deficiency being made up by the injecting power.

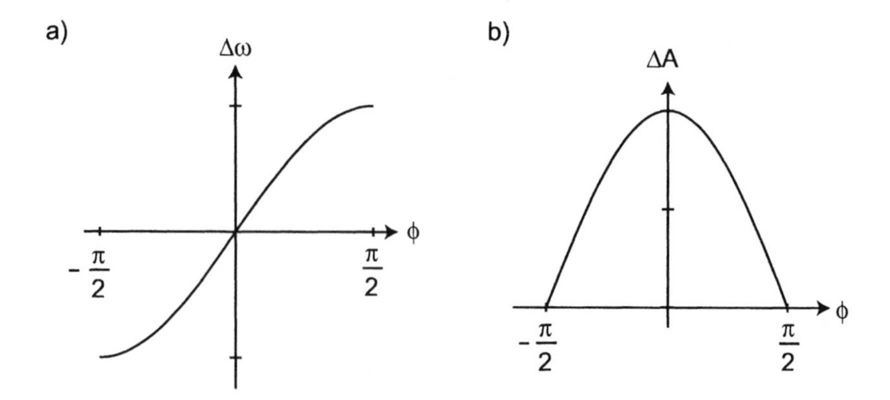

Figure 7.12: Injection locking range. Left-hand trace: the detuning $\Delta\omega$ is the difference between the injection signal frequency and the free-running oscillator frequency. The angle ϕ is the phase between two signals. It has a range of $\pm\pi/2$. When the injection is successful, the oscillator oscillates at the injection frequency and has a well-defined phase ϕ relative to the injection signal. Right-hand trace: increased current amplitude due to injection. It is maximum when $\phi = 0\,(\Delta\omega = 0)$, and is zero at the limits of the locking range, $\phi = \pm\pi/2$.

7.B.4 Noise-perturbed oscillator

When the injection signal represents noise and fluctuates randomly in time, Eq. 7.41 can be solved only in a statistical sense. In particular, we are interested in the phase fluctuation, which determines the ultimate finite frequency width of an oscillator (or laser). The amplitude fluctuations are damped, as can be seen from Fig. 7.11, since a decrease in the amplitude, for example, leads to an increase in the gain that restores the amplitude. The noise source fluctuates, but within a narrow frequency band near the oscillation frequency. In the phase equation, the term $d\phi/dt$ consists of a possible constant part that goes into the steady-state oscillation frequency; the fluctuating part of the phase is governed by the following equation:

$$\frac{d\phi}{dt} = \frac{1}{2LA_0}\text{Im}\left[v_s(t)e^{-i\omega t}\right]$$

Note that the driving term within the square brackets varies slowly, its center frequency being cancelled by the factor $\exp(-i\omega t)$. This equation describes random walk.

Before proceeding to solve the equation, some mathematical preliminaries are in order. We denote

$$\frac{1}{2LA}\text{Im}\left[v_s(t)e^{-i\omega t}\right]$$

by $V_{\text{sq}}(t)$. The term V_{sq} varies randomly. If we multiply $V_{\text{sq}}(t)$ by $V_{\text{sq}}(t')$, and average over many such products, we expect the average to vanish, since any one product is equally likely to be positive or negative, except when $t = t'$, in which case we get the average of the magnitude squared. This argument leads to

$$\langle V_{sq}(t)V_{sq}(t')\rangle = D\delta(t-t')$$

where D is spectral power density of V_{sq}, as a dimensional analysis or a straight forward calculation of the Fourier transform shows. The quantity on the left-hand side is the correlation function of V_{sq}, and as it depends only on the difference of the two times t and t' and not on t and t' separately, we can rewrite it in terms of the time difference τ:

$$\langle V_{sq}(\tau)V_{sq}(0)\rangle = D\delta(\tau)$$

Similarly, the spectrum of the current is the Fourier transform of its correlation function

$$\langle I(\tau)I(0)\rangle$$

If we ignore the small, well-damped amplitude fluctuations, then

$$\langle I(\tau)I(0)\rangle \simeq A^2\left\{\exp(-i\omega_0\tau)\right\}\langle \exp -i[\phi(\tau)-\phi(0)]\rangle$$

The calculation of the angled-bracketed quantity is lengthy.[11] It is performed by expanding the exponential. In the expansion, products with an odd number of terms (which are purely imaginary) average to zero, and the resummation of the products with an even number of terms (which are real) leads to a simple result:

$$\langle \exp -i[\phi(\tau)-\phi(0)]\rangle = \exp\left[-\frac{1}{2}\left\langle |\phi(\tau)-\phi(0)|^2\right\rangle\right]$$

The quantity in the exponent on the right-hand side can be calculated by directly integrating Eq. 7.41 and performing the statistical average:

$$\begin{aligned}
\int_0^\tau \frac{d\phi}{dt}dt &= \int_0^\tau dt V_{sq}(t), \\
\left\langle |\phi(\tau)-\phi(0)|^2\right\rangle &= \int_0^\tau dt' \int_0^\tau dt < V_{sq}(t)V_{sq}(t') > \\
&= \int_0^\tau dt' \int_0^\tau dt\ D\delta(t-t') \\
&= D\,|\tau|
\end{aligned}$$

and hence

$$\langle I(\tau)I(0)\rangle \simeq A^2\left\{\exp(-i\omega_0\tau)\right\}\exp\left(-\frac{D}{2}|\tau|\right)$$

[11]See M. Sargent III, M. O. Scully, and W. E. Lamb, Jr., *Laser Physics*, Addison-Wesley, Reading, MA, 1974

We can now obtain the spectrum of $I(t)$ by the inverse Fourier transform

$$\begin{aligned}\langle I(\Omega)^2\rangle &= \int d\tau\,\langle I(\tau)I^*(0)\rangle \exp(i\Omega\tau)\\ &= A^2\int d\tau \exp\left(-\frac{D}{2}|\tau| - i\omega_0\tau + i\Omega\tau\right)\\ &= \frac{A^2}{(\Omega-\omega_0)^2+(D/2)^2}\end{aligned}$$

The spectrum is a Lorentzian centered on the unperturbed frequency ω_0 with a full width of D. To use this result in more general cases than the particular $L-C$ resonance circuit here, let us recast D in more general physical quantities.

7.B.5 Oscillator line width and the Schawlow–Townes formula

By definition, we have

$$\begin{aligned}D\delta(\tau) &= \langle V_{\text{sq}}(\tau)V_{\text{sq}}(0)\rangle\\ &= \left(\frac{1}{2LA_0}\right)^2 \langle \text{Im}\, v_s(\tau)e^{-i\omega\tau}\text{Im}\, v_s(0)\rangle\end{aligned}$$

Fourier transforming this equation yields

$$D = \left(\frac{1}{2LA_0}\right)^2 \langle |\text{Im}\, v_s(\Omega)|^2\rangle$$

where the angle-bracketed quantity is the in-quadrature noise voltage squared per unit frequency, a constant for white noise assumed here ("white" over the resonator width). Manipulating the right-hand side as in the case of injection locking, we can rewrite D as

$$D = \Delta\omega_e^2 \frac{P_{\text{noise}}(\Omega)}{P_{\text{output}}} \tag{7.42}$$

where $P_{\text{noise}}(\Omega)$ is the noise power per unit frequency into the load resistor R. Note the dimension of $P_{\text{noise}}(\Omega)$ is energy. This is a very important and general result that deserves further discussion. The line width of an oscillator is not the passive ("cold") resonator width $\Delta\omega_e$; rather, it is that width reduced by the ratio $\Delta\omega_e P_{\text{noise}}(\Omega)/P_{\text{output}}$. Since $\Delta\omega_e P_{\text{noise}}(\Omega)$ is the total noise power into the resonator bandwidth, the ratio is simply noise power over signal power.

In a resonator with many resonances or modes, $P_{\text{noise}}(\Omega)$ is the noise power per unit frequency into *one* mode. In a laser, the fundamental, unavoidable noise source is spontaneous emission, which accompanies any stimulated emission. In this case, $P_{\text{noise}}(\Omega)$ is particularly simple; it is equal to one photon energy $\hbar\omega$. This is from a well-known result from statistical mechanics, derived heuristically below, which says the spontaneous emission into one mode is one photon per unit bandwidth. Substitution into Eq. 7.42 yields the Schawlow–Townes formula for the fundamental line width of a laser:

$$D = \Delta\omega_e^2 \frac{\hbar\omega}{P_{\text{output}}}$$

In almost all lasers, the line widths are limited by extraneous factors such as mirror vibration, thermal fluctuation of the refractive index of the medium. The only laser whose line width is in reality limited by spontaneous emission noise is the semiconductor laser, because of its large $\Delta\omega_e$ and small P_{output}. In passing, we note that the Schawlow–Townes formula is the high-frequency version of the microwave oscillator line width limited by thermal radiation, where the unit of thermal energy kT replaces $\hbar\omega$.

We now calculate $P_{\text{noise}}(\Omega)$ into one mode of a laser resonator. The laser resonator is the subject of Chapter 8, from which we use some results. Consider a laser resonator consisting of two mirrors separated by a distance L. For simplicity, we assume the lasing atoms completely fill the space between the mirrors. The oscillating mode has a cross-sectional area A. If there are N_2 atoms per unit volume in the upper state, the *total* radiation power emitted by spontaneous emission is

$$(N_2AL)\hbar\omega\gamma$$

where γ is the spontaneous emission rate. This power is distributed in frequency and in space. In frequency, it is distributed over the line width Γ, but we are limited to one axial mode of width $c/2L$. In space, the power is radiated over the full solid angle 4π . The solid angle subtended by the mode is approximately $(\lambda/\sqrt{A})^2 = \lambda^2/A$, where $\lambda/\sqrt{A}$ is the diffraction angle of the mode beam. The fraction of power radiated into the mode is therefore $(\lambda^2/A)/(4\pi)$. Moreover, the radiation can be in either of the two independent polarizations, so the power radiated by spontaneous emission into one mode is

$$\begin{aligned} P_{\text{noise}}(\Omega) &= (N_2AL)\,\hbar\omega\gamma \times \frac{c/2L}{\Gamma} \times \frac{\lambda^2/A}{4\pi} \times \frac{1}{2} \\ &= cN_2\left(\frac{1}{16\pi}\lambda^2\frac{\gamma}{\Gamma}\right)\hbar\omega. \end{aligned}$$

The factor within parentheses is, to within a numerical factor, the stimulated emission cross section σ. So

$$\begin{aligned} P_{\text{noise}}(\Omega) &\simeq cN_2\sigma\hbar\omega \\ &= \Delta\omega_e\hbar\omega. \end{aligned}$$

where the last line follows from the fact that $cN_2\sigma$ is the gain rate, which must be equal to the loss rate $\Delta\omega_e$.

Problem 7.6 *Use the Schawlow-Townes formula to estimate the intrinsic line width of a semiconductor laser and a dye laser. Data on the semiconductor are in the example discussed in Section 7.4.3 . For the dye laser, assume a resonator length of 1 m, 5% output mirror coupling, and 50 mW output power.*

Chapter 8

Elements of Optics

8.1 Introduction

So far, we have treated light as a plane wave in its interaction with atoms. This is an excellent approximation because the characteristic atomic length scale is much smaller than an optical wavelength, and the light field is uniform across the extent of the atom. In the laboratory, however, we manipulate light on a scale much larger than that of a wavelength. To do this effectively, we need to understand the propagation of light beyond the plane wave approximation. This brings us to the subject of optics. In this chapter, we restrict ourselves to light propagating in homogeneous media, thus leaving out the important subject of waveguides. We begin with *geometric* or *ray* optics, and introduce the $ABCD$ matrices, which are a convenient way to follow the propagation of light. Ray optics is the limiting case of very short wavelengths. The defining features of waves, interference and diffraction, are lost in ray optics. In an analogy with mechanics, ray optics is like the classical limit of particle mechanics. Unlike the wavefunction of quantum mechanics, which applies only at atomic scales and below, the wave nature of light is more readily observed because optical wavelengths, on the order of a micrometer, are usually much longer than de Broglie wavelengths of particles. In many circumstances, therefore, a more accurate, *wave* description of light is needed, which is developed after geometric optics. Both theories developed here, geometric and wave, apply only to light rays and waves of small divergence angles, such as beams from lasers, or light from a distant source. Furthermore, the wave field is *scalar*, whereas the electric field of a light wave is a vector. However, close to the axis of a laser beam, the polarization is almost uniform with only one component for which the scalar field is a good approximation. Once the scalar field is found, we will show how to find the vector fields from the scalar field. Our treatment combines the traditional Fresnel diffraction theory and Gaussian beams of laser optics. We consider Gaussian beams as a special case of the general Fresnel diffraction theory. Finally, the $ABCD$ matrices are applied to the Gaussian beams for the design of laser resonators.

In our presentation of diffraction theory, we value mathematical simplicity above all else. In particular, since all the important concepts can be understood with waves in two-dimensional space, we will develop a detailed theory in only two dimensions. Corresponding results for the more physical three-dimensional waves are stated with no proof or with proof briefly sketched. Moreover, only monochromatic waves are considered. Most optical experiments are now performed with laser light, for which a monochromatic wave is a good approximation in most cases. To apply the theory to incoherent light, simply add the intensities of each frequency component at the end.

8.2 Geometric Optics

Traditional optical designs begin with paraxial geometric optics, followed by necessary corrections and refinements. In manipulation of laser light, often a knowledge of wave optics is required. Still, paraxial geometric optics provides quick and intuitive, and often correct, answers. We expect the readers of this text to have had some previous exposure to geometric optics which will serve as a conduit to many concepts and techniques. We can only touch on the bare essentials of geometric optics in this section; design of optical components is a subject in itself and must be left to specialized books. As geometric optics is a limiting case of wave optics, logically one would expect to study wave optics first, then take the short wavelength limit to geometric optics. This approach turns out to be a rather difficult mathematical problem. Instead, we illustrate their relationship by working out several examples in both geometric and wave optics.

When light is considered as rays propagating along an axis through a line of optical elements, each ray is completely characterized by two parameters: the distance r of the ray from the axis, and the angle r' the ray makes with the axis at that point (Fig. 8.1). An optical element may be a lens, a mirror, or simply a stretch of space or material spanning a distance d. We only consider small angles. This is not a severe restriction in practice. For incoherent sources like lamps, most of the time we intercept light only at a distant point; laser beams usually have small divergence angles. The parameters (r_2, r_2') of a ray leaving any of the optical elements considered here are related linearly to those (r_1, r_1'). Since a linear relationship can be represented by a matrix, we write

$$\begin{pmatrix} r_2 \\ r_2' \end{pmatrix} = \begin{pmatrix} A & B \\ C & D \end{pmatrix} \begin{pmatrix} r_1 \\ r_1' \end{pmatrix} \tag{8.1}$$

The transmission of a ray through an optical element is completely characterized by the $ABCD$ matrix of the element. It is remarkable that the same matrix applies to the propagation of a Gaussian beam, for two different parameters which characterize the Gaussian beam. Note the dimensions of the matrix elements: A and D are dimensionless; C has the dimension of length, while D has the dimensions of inverse length. For our purposes, we need the matrices of

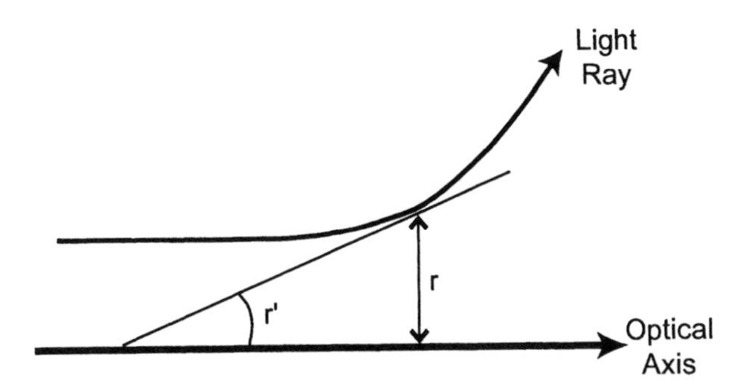

Figure 8.1: Ray displacement and slope. A light ray is characterized by its displacement from the optical axis and the slope of the ray at that point.

only two elements: homogeneous material of length d, and a thin lens of focal length f.

8.2.1 *ABCD* matrices

We now derive the *ABCD* matrices for the two optical elements mentioned above. For rays going through a distance d in a homogeneous medium, consider a ray displaced by r_1 from the optical axis and propagating at a slope r'_1. After a distance d, the displacement becomes $r_1 + dr'_1$, and the slope remains r'_1, that is, $r_2 = r_1 + dr'_1$ and $r'_2 = r'_1$. Or, the *ABCD* matrix for homogeneous material of length d is

$$\begin{pmatrix} A & B \\ C & D \end{pmatrix} = \begin{pmatrix} 1 & d \\ 0 & 1 \end{pmatrix} \equiv \mathrm{M}_d \tag{8.2}$$

where the matrix has been identified by the subscript d. For a thin lens of focal length f, we can follow a ray as it enters, propagates inside, and exits the lens, and relate the exit (r, r') parameter to the entrance. However, it is simpler to use the focusing property expected of a thin lens to derive the *ABCD* matrix. First, "thin" means that the ray is not displaced by traveling through the lens. We expect rays parallel to the optical axis entering the lens $(r_1, r'_1 = 0)$ to be focused to a point at a distance f away from the lens on the axis, $(r_2 = r_1, r'_2 = -r_1/f)$. Hence $A = 1$, $C = -1/f$. Similarly, rays passing through the optical axis at a distance f in front of the lens, $(r_1, r'_1 = r_1/f)$, are collimated $(r_2 = r_1, r'_2 = 0)$. Hence $B = 0$ and $D = 1$. The *ABCD* matrix for a thin lens is therefore

$$\begin{pmatrix} A & B \\ C & D \end{pmatrix} = \begin{pmatrix} 1 & 0 \\ -1/f & 1 \end{pmatrix} \equiv \mathrm{M}_f \tag{8.3}$$

where the matrix has been identified by the subscript f. A lens with negative f is a divergent lens. Let us illustrate the application of the matrices by applying

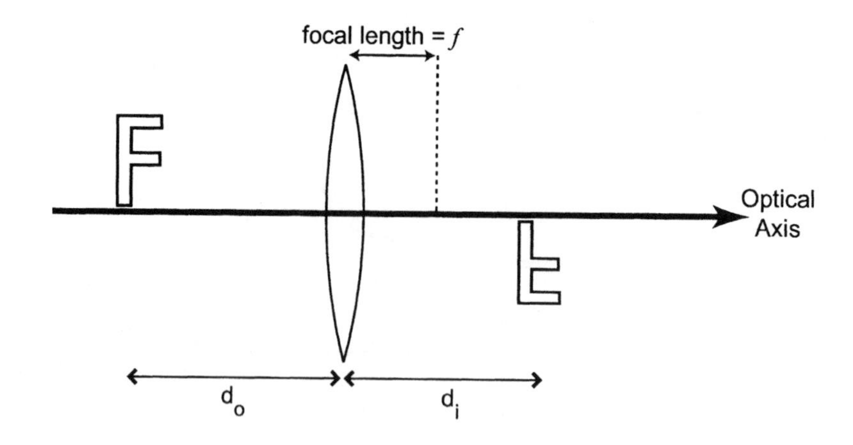

Figure 8.2: Imaging by a thin lens. An image of an object formed by a thin lens at a distance d_i. The object is shown at a distance d_o from the lens, where $d_o > f > 0$. The image is real and inverted.

them to a familiar problem.

Example 8.1 Imaging by a single thin lens

An object of height h is located at a distance d_o to the left of a lens of focal length f (Fig. 8.2). Let us find the location and size of the image. An image is an exact replica of the object, with the possibility of magnification or rotation. The light rays leaving a point of the object at any angle must converge on the same point at the image. This fact allows us to find the image. A ray characterized by (r_1, r_1') from a point on object is transformed to $\begin{pmatrix} r_2 \\ r_2' \end{pmatrix} = M_{d_o} \begin{pmatrix} r_1 \\ r_1' \end{pmatrix}$ in front of the lens. The lens transforms the ray to $\begin{pmatrix} r_3 \\ r_{3'} \end{pmatrix} = M_f \begin{pmatrix} r_2 \\ r_{2'} \end{pmatrix}$. At the image a distance d_i from the lens, the ray is transformed to $\begin{pmatrix} r_4 \\ r_{4'} \end{pmatrix} = M_{d_i} \begin{pmatrix} r_3 \\ r_{3'} \end{pmatrix}$. Therefore

$$\begin{pmatrix} r_4 \\ r_{4'} \end{pmatrix} = \mathsf{M}_{d_i}\mathsf{M}_f\mathsf{M}_{d_o} \begin{pmatrix} r_1 \\ r_{1'} \end{pmatrix}.$$

This equation illustrates a very important fact: *In passing through a series of optical elements, a ray is transformed by a matrix that is the product of the ABCD matrices of the individual elements, the order of the matrices is in the reverse order of the elements.* In this example,

$$\mathsf{M}_{d_i}\mathsf{M}_f\mathsf{M}_{d_o} = \begin{pmatrix} 1 - \frac{d_i}{f} & d_o\left(1 - \frac{d_i}{f}\right) + d_i \\ -\frac{1}{f} & 1 - \frac{d_o}{f} \end{pmatrix}$$

Now all rays, regardless of initial slope from the object, must arrive at the same point at the image; and thus r_4 must be independent of r_1'. Therefore the element C of the matrix product must be zero: $d_o(1 - d_i/f) + d_i = 0$, or by dividing the equation by $d_i d_o$, we get the familiar lens equation:

$$\frac{1}{d_o} + \frac{1}{d_i} = \frac{1}{f} \tag{8.4}$$

which yields the position of the image. With this equation, the matrix for the imaging lens above can be simplified to

$$\begin{pmatrix} -\frac{d_i}{d_o} & 0 \\ -\frac{1}{f} & -\frac{d_i}{d_0} \end{pmatrix}.$$

Applying this matrix, we obtain

$$r_4 = -\left(\frac{d_i}{d_o}\right) r_1$$

which means, physically, that the object is magnified by the factor d_i/d_o. By definition, d_o is positive. If f is positive and the object is outside the focal length, $(d_o > f)$ then $d_i > 0$ and $r_4/r_1 < 0$, which means that a real, inverted image is formed on the opposite side of the lens from the object. If $d_o < f$, then the solution for d_i from the lens equation is negative, and $r_4/r_1 > 0$. The image is on the same side of the object and is upright. However, the image is virtual, as the rays do not converge on, but only appear to emanate from, the image. The common handheld magnifying glass is used in this way for reading, with $f \simeq 8$ in. (20.3 cm) and $d_o \simeq 3$ in. (7.62 cm), resulting in a magnification of about 150%. For $f < 0$, the image is always virtual, upright, and on the same side as the object.

One limitation of geometric optics can be pointed out immediately in this example. Suppose that the object is at infinity. The incident rays are then essentially parallel. Geometric optics predicts a zero image size. The actual image size can only be calculated using wave optics.

This is an appropriate place for a few remarks on the imperfections of optical systems ignored in paraxial optics, but with which traditional optical engineers must deal. Take the single lens as an example. We used the focusing properties of the a single lens to derive its $ABCD$ matrix. The only parameter characterizing the lens is its focal length, irrespective of the lens shape. In reality, lenses of the same focal length but different shapes can have very different characteristics. For example, a double convex lens focuses a beam better than other shapes. In professional parlance, the double convex lens has the least spherical aberration. And in using a planoconvex lens to focus a beam, better focus is achieved by orienting the convex face to the incident parallel beam than orienting the plane face. Of course, there are considerations other than spherical aberration. For example, despite its largest spherical aberration of all lens shapes, the meniscus, a lens formed by two convex or two concave surfaces, is used for spectacles, because at the large angles in normal vision, the image distortion is least. Mirrors perform similar functions as lenses. The choice between them depends on the application. For example, the focal lengths of mirrors do not depend on wavelengths as much as those of lenses (a property called *chromatic aberration*), but mirrors return incident beams to the incident direction, which can be inconvenient in some applications. A ray incident on a mirror is totally reflected at an angle equal to the incidence angle (Fig. 8.3a). Rays parallel to the optical axis incident on the mirror at small angles converge on the optical axis half way between the center and the apex of the spherical mirror surface (Fig. 8.3b). The focal length of the mirror is therefore equal to half of the mirror radius. In optical systems containing mirrors, instead of following the actual direction of the rays on each reflection and therefore changing the direction of the optical axis, we usually trace the rays in one direction only, a practice called "unfolding" the system. The reflection from a mirror then is equivalent to the transmission through a lens of focal length equal to half of the mirror radius. Large astronomical telescopes use mirrors for reasons of weight, size, and chromatic aberration.

Example 8.2 The compound lens

A compound lens consisting of two simple lenses has much greater flexibility and functionality than does a single simple lens. Compound lenses are used widely to compensate for lens aberrations, a subject beyond this book. Two simple lenses together solves many problems impractical for one. For example, when a large magnification is needed, even with a perfect lens, the image distance may be impractically large. By changing the ratio of the focal lengths and their ratio to the separation between them, one can design different instruments like the microscope, the telescope, and the beam expander. We discuss the compound lens in this example to illustrate the use of *ABCD* matrices and to introduce some common terms in optics (principal planes, effective focal length).

Consider the system of two lenses, of focal lengths f_1 and f_2, separated by a distance d (Fig. 8.4). We find the *ABCD* matrix for the system by multiplying the matrices for the lenses and the space as follows:

$$\begin{pmatrix} 1 & 0 \\ -\frac{1}{f_2} & 1 \end{pmatrix} \begin{pmatrix} 1 & d \\ 0 & 1 \end{pmatrix} \begin{pmatrix} 1 & 0 \\ -\frac{1}{f_1} & 1 \end{pmatrix}$$

$$= \begin{pmatrix} 1 - \frac{d}{f_1} & d \\ -\frac{(f_1+f_2-d)}{f_1 f_2} & 1 - \frac{d}{f_2} \end{pmatrix}$$

$$\equiv \begin{pmatrix} 1 - \frac{d}{f_1} & d \\ -\frac{1}{f_{\mathrm{eff}}} & 1 - \frac{d}{f_2} \end{pmatrix} \tag{8.5}$$

$$\equiv \begin{pmatrix} A & B \\ C & D \end{pmatrix}_{\mathrm{cpd}} \tag{8.6}$$

where we have defined an effective focal length f_{eff} as $f_1 f_2/(f_1 + f_2 - d)$. The reason will become clear immediately. Suppose that a ray parallel to, and at a

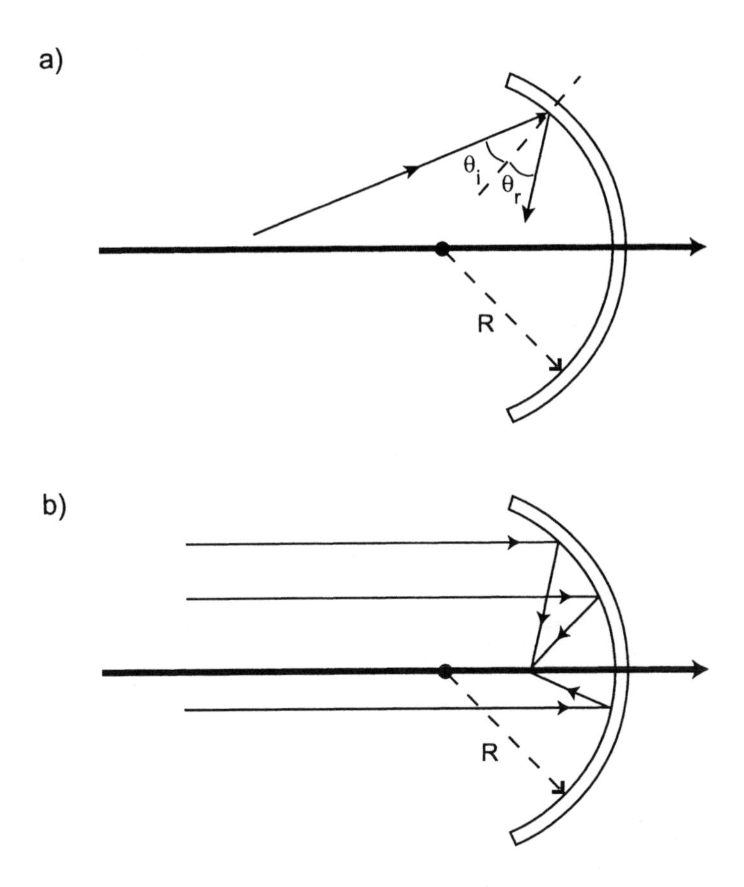

Figure 8.3: The spherical mirror. (a) A ray entering the mirror at an angle θ_i relative to the mirror normal is reflected back at an angle $\theta_r = \theta_i$. (b) The optical axis is normal to the mirror surface at the point of intersection. Rays parallel and close to the optical axis are all reflected to pass through the axis at a point that is half a radius from the mirror. The focal length of a mirror is therefore half of its radius.

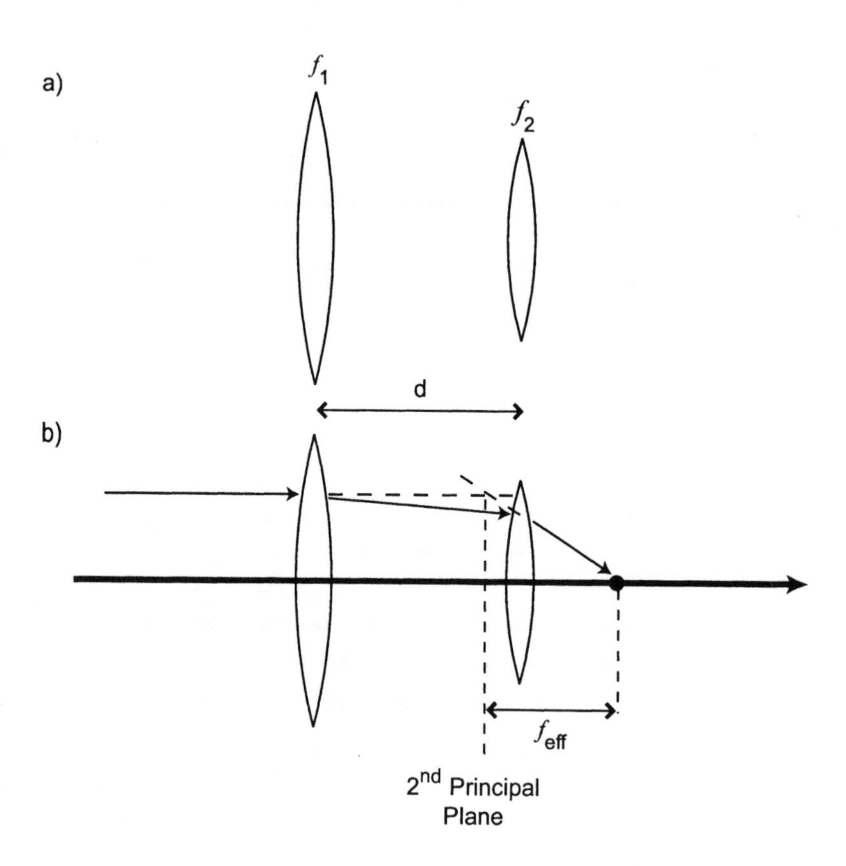

Figure 8.4: The compound lens. (a) A two-element compound lens. (b) Rays parallel to the optical axis are focused to a point that is f_{eff} from the second principal plane. The principal plane is where the extended incident ray and the extended exit ray meet.

distance r from, the optical axis enters from the left. The exit distance and angle are given by

$$\begin{pmatrix} A & B \\ C & D \end{pmatrix}_{\text{cpd}} \begin{pmatrix} r \\ 0 \end{pmatrix} = \begin{pmatrix} \left(1 - \frac{d}{f_1}\right) r \\ -\frac{1}{f_{\text{eff}}} r \end{pmatrix}$$

The exit ray intercepts the optical axis at a distance $f_{\text{eff}}(1 - d/f_1)$, independent of the initial ray distance r (Fig. 8.4b). Hence a bundle of parallel rays from the left will all converge on that point, which is called the *second focal point* of the compound lens. Now, if we extend the exit ray back towards the lens, and extend the incident ray, the two will intercept at a distance f_{eff} from the second focal point. The plane perpendicular to the optical axis and passing through this intercept is called the *second principal plane*.

Similarly, a ray parallel to the optical axis entering from the right intercepts the optical axis at a distance $f_{\text{eff}}(1 - d/f_2)$ from the front of the lens, called the *first focal point*. The plane passing through the extended incident and exit rays is called the *first principal plane*.

The simple lens formula Eq. 8.4 applies to the compound lens, if the object distance is measured from the first principal plane, the image distance measured from the second principal plane, and the focal length replaced by the effective focal length.

Example 8.3 The beam expander

The beam expander enlarges a parallel beam of radius a to a parallel beam of radius b. It requires at least two lenses since one single lens will focus the incident parallel beam which subsequently diverges. Let the focal lengths of the lenses be f_1 and f_2, and the distance between them d. The results from the example of compound lens above, Example 8.2, can be applied immediately:

$$\begin{aligned} \begin{pmatrix} A & B \\ C & D \end{pmatrix}_{\text{cpd}} \begin{pmatrix} a \\ 0 \end{pmatrix} &= \begin{pmatrix} \left(1 - \frac{d}{f_1}\right) a \\ -\frac{a}{f_{\text{eff}}} \end{pmatrix} \\ &= \begin{pmatrix} b \\ 0 \end{pmatrix} \end{aligned}$$

Thus

$$b = a\left(1 - \frac{d}{f_1}\right)$$

and $a/f_{\text{eff}} = 0$ or

$$f_1 + f_2 = d$$

These two conditions lead to $f_2/f_1 = -b/a$. The expander is illustrated in Fig. 8.5a. There is another solution. If we set the ray output to be $(-b, 0)$ instead, then $f_2/f_1 = +b/a$. This arrangement is shown in Fig. 8.5b.

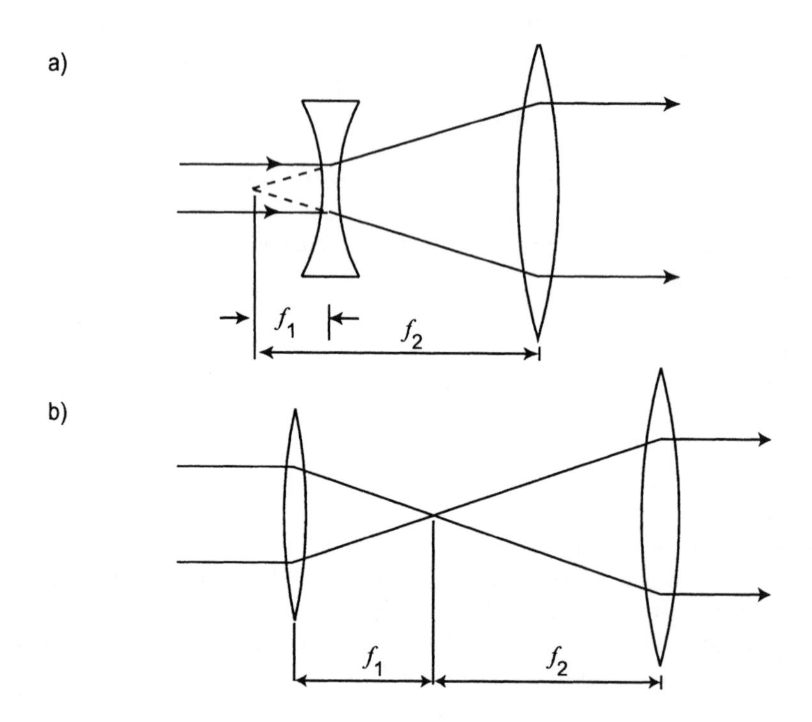

Figure 8.5: The two-element beam expander. (a) Expander with one positive lens (f_2) and one negative lens (f_1). The distance between the lenses is $f_2 - f_1$. The incident beam is expanded by the negative lens and diverges. The positive lens refocuses the beam. (b) Expander with two positive lenses. The distance between the lenses is equal to the sum of the focal lengths. The entering beam is focused by the first lens, diverges, and is then refocused by the second lens.

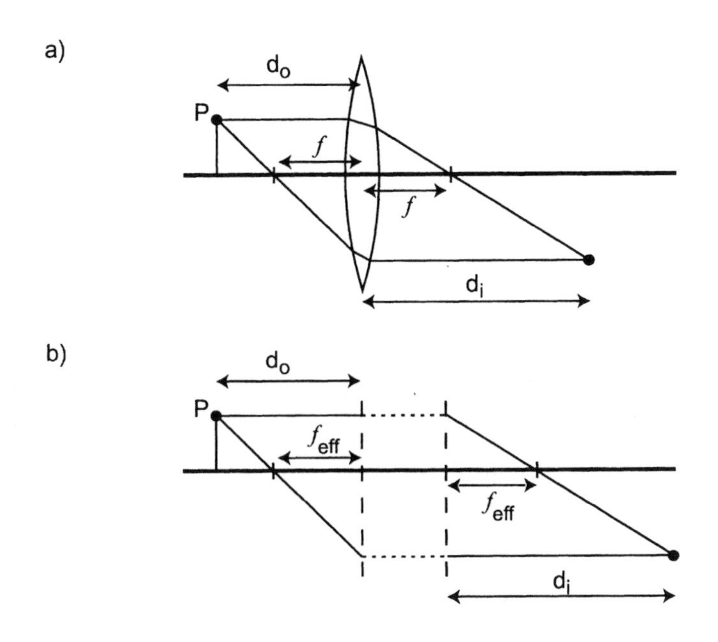

Figure 8.6: Imaging by a single lens (a) and compound lens (b).

Problem 8.1 *The lens law can be derived using ray tracing instead of ABCD matrices.*

(a) Simple lens

(i) Refer to Fig. 8.6a. Consider two rays from a point P of the object at a distance d_o, one parallel to the optical axis, the other passing through the focal point. After going through the lens, the first ray must pass the focal point at the other side of the lens, and the second ray must be parallel to the optical axis. The image of P is the point at which these two rays meet. Prove that the distance d_i of the image from the lens is given by the lens law.

(ii) Figure 8.6a represents the case $d_o > f > 0$. Convince yourself that the lens law holds for the other cases ($f > d_o > 0$, $f < 0$) as well.

(b) Compound lens

Use the same technique to prove that the lens law holds for a compound lens, provided d_o and d_i are measured from the first and second principal planes, and f_{eff} is used for the focal length (Fig. 8.6b).

We conclude geometric optics with a discussion of ray transmission through a dielectric interface, which is governed by Snell's law and is the starting point of calculating the transmission through many optical elements such as lenses, prisms, and waveguides. In general, when light comes to a dielectric interface, it will be partially reflected and partially transmitted. We will ignore the reflec-

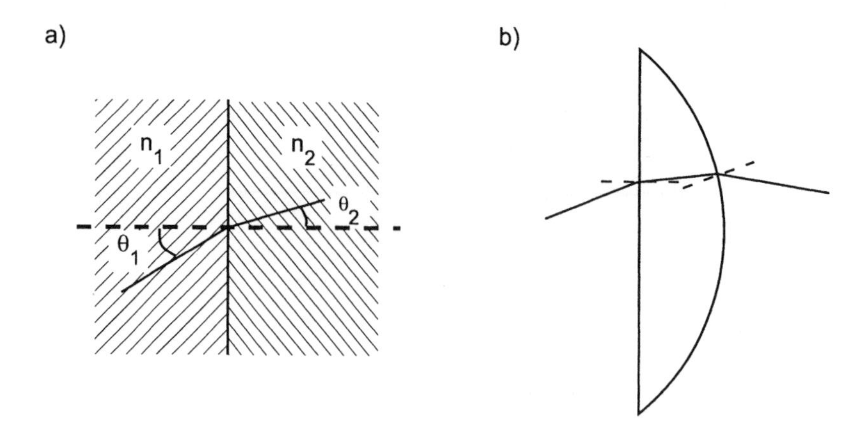

Figure 8.7: Dielectric interface and planoconvex lens. (a) A ray entering from one dielectric to another is deflected. With angles measured relative to the normal of the interface, Snell's law leads to the relationship $n_1\theta_1 = n_2\theta_2$, provided the angles are small. (b) The path of a ray through a thin lens is deflected twice, one at entrance and once at exit. Each time, the deflection depends on the incident angle relative to the normal of the interface at the point of incidence. For a planoconvex lens with one surface of radius R, the focal length is $f = R/(n-1)$.

tion. Consider a ray entering an interface between two dielectric media. The interface need not be planar. The ray enters the interface from medium 1 at an angle θ_1 from the normal to the interface, and is transmitted to the second medium at angle θ_2 to the normal (Fig. 8.7a).

By Snell's law,

$$n_1 \sin\theta_1 = n_2 \sin\theta_2$$

where n_1 and n_2 are the refractive indices of medium 1 and medium 2. At small angles

$$n_1\theta_1 \simeq n_2\theta_2$$

but $r_1^{'} \simeq \theta_1$, $r_2^{'} \simeq \theta_2$, so

$$n_1 r_1^{'} = n_2 r_2^{'}$$

In applying this equation to find the transmission through a curved interface, the surface of a lens, for example, we have to consider the incident ray at an arbitrary point, and the normal to the surface at that point is the local optical axis, which must then be related to the axis of the lens. Figure 8.7b shows a ray passing through a planoconvex thin lens one of whose surfaces is plane and the other a sphere of radius R. The term "thin" means the change in the distance r inside the lens can be ignored. Applying the formula above to trace the ray through the lens, and comparing with the $ABCD$ matrix for a thin lens, the

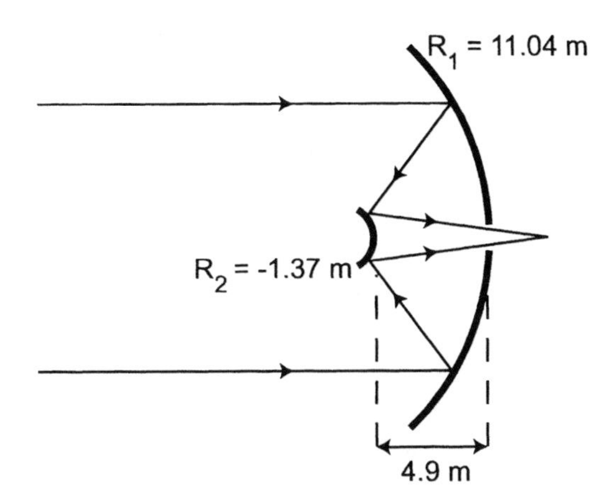

Figure 8.8: Optics of the Hubble Space Telescope

focal length f of the lens can be found to be

$$\frac{1}{f} = \frac{n-1}{R}$$

Problem 8.2 *The Hubble Space Telescope:*

The optics of the Hubble Space Telescope (HST) (see Fig. 8.8) consists of a concave mirror of radius of curvature 11.04 m and a convex mirror with radius of curvature 1.37 m. The mirrors are separated by 4.9 m. The diameter of the concave mirror is 2.4 m.

(a) What is the equivalent lens system of the HST?

(b) What is the ABCD matrix for the lens equivalent of the HST? The input is right in front of the first lens, and the output is right after the second lens.

(c) An object of height h_o is at a large distance d_o from the HST. Use the ABCD matrix to find (i) the image distance d_i from the second lens; (ii) the image height h_i; (iii) the angular magnification or power $(h_i/d_i)/(h_o/d_o)$. You will find that the power is not very different from that of an inexpensive telescope. What do you think makes the HST such a valuable instrument?

(d) The closest distance of the planet Jupiter to the earth is 45 million miles (72 million km). What is the diameter of the smallest area on Jupiter that HST can resolve, and what is the size of the image of that area? The diameter of Jupiter is 86,000 miles (138,000 km). What is the size of the image of Jupiter?

(e) Wouldn't the front mirror and the hole in the back mirror make a dark spot in the center of the image?

8.3 Wave Optics

8.3.1 General concepts and definitions in wave propagation

We consider monochromatic waves of frequency ω. Other waveforms in time can be synthesized by superpositions of waves of many frequencies. We also restrict any detailed discussions to *scalar* waves. Light waves of electric and magnetic fields are vectors, but near the center of an optical beam the fields are very nearly uniformly polarized and a scalar wave representing the magnitude of the field is a very good approximation. The field $\psi(\mathbf{r}, t)$ is governed by the scalar wave equation

$$\nabla^2 \psi = \frac{1}{c^2} \frac{\partial^2 \psi}{\partial t^2} \tag{8.7}$$

Let ψ be of the form

$$\psi(\mathbf{r}, t) = A(\mathbf{r}) \exp\{i\,[\phi(\mathbf{r}) - \omega t]\}$$

where A and ϕ are real functions of space. A is the *amplitude* of the wave, and the exponent within the square brackets is called the *phase* of the wave. Sometimes, where the context is clear, we also call ϕ the phase. In this form, it is implicit that the rapid variations in space and in time are contained in the phase. The surface obtained by setting the phase equal to a constant

$$\phi(\mathbf{r}) - \omega t = \text{constant}$$

is called a *wavefront* or *phase front*. Since there are infinitely many possible constants, there are infinitely many possible wavefronts. The rapid motion associated with a wave can be followed by following the motion of a particular wavefront. The interference pattern between two waves is largely formed by the wavefronts of the two waves. The velocity at which a particular wavefront or phase front moves is called the *phase velocity*. Suppose that we follow a particular wavefront at time t. At time $t + \Delta t$, the wavefront will move to another surface. A point $\mathbf{r}$ on the original surface will move to another point $\mathbf{r} + \Delta\mathbf{r}$ (Fig. 8.9):

$$\phi(\mathbf{r} + \Delta\mathbf{r}) - \omega(t + \Delta t) = \phi(\mathbf{r}) - \omega t = \text{constant}$$

Expanding $\phi(\mathbf{r} + \Delta\mathbf{r}) \simeq \phi(\mathbf{r}) + \nabla\phi(\mathbf{r}) \cdot \Delta\mathbf{r}$, we obtain

$$\nabla\phi(\mathbf{r}) \cdot \Delta\mathbf{r} = \omega \Delta t$$

$\nabla\phi(\mathbf{r})$ is normal to the phase front, and is called the *wave vector*. The term $\Delta\mathbf{r}$ is smallest if it lies in the direction of $\nabla\phi$, and the wavefront travels with the speed

$$\frac{|\Delta\mathbf{r}|}{\Delta t} = \frac{\omega}{|\nabla\phi(\mathbf{r})|}$$

which is the phase velocity. The phase velocity can vary from point to point in space. We now look at a few examples.

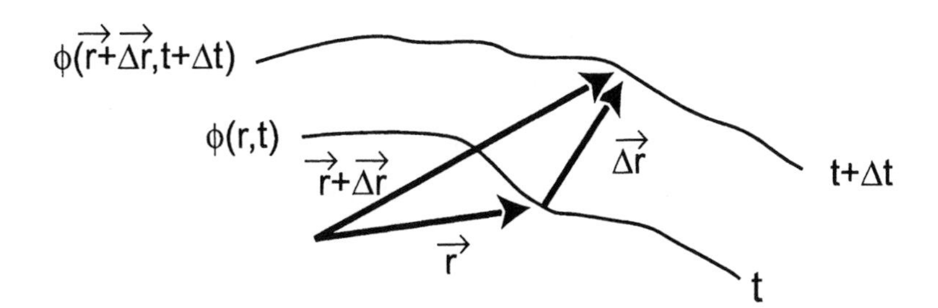

Figure 8.9: Propagation of a wavefront. Wavefronts are surfaces of constant phase. Shown are a wavefront at time t and at another time $t + \Delta t$. A point on the wavefront at t moves to another point on the wavefront at $t + \Delta t$, the displacement between these two points is $\Delta\mathbf{r}$.

Example 8.4 Plane wave

Begin with the solution to the scalar wave equation, Eq. 8.7

$$\psi(\mathbf{r}, t) = A_0 \exp[i\mathbf{k} \cdot \mathbf{r} - i\omega t]$$

Substituting ψ into the scalar wave equation (Eq. 8.7) yields $|\mathbf{k}| = \omega/c$ when A_0 is constant. The wavefront is defined by

$$\mathbf{k} \cdot \mathbf{r} - \omega t = \text{constant}$$

or, at a given t, $\mathbf{k} \cdot \mathbf{r} = \omega t + \text{constant}$ which is a plane perpendicular to the vector $\mathbf{k}$. Since $\phi = \mathbf{k} \cdot \mathbf{r}$, $\nabla\phi = \mathbf{k}$, and the phase velocity $\omega / |\nabla\phi| = c$. Along the direction of $\mathbf{k}$, the wave is periodic in space. The period, called the wavelength λ, is given by $|\mathbf{k}| = 2\pi/\lambda$.

Example 8.5 Spherical wave

We write the expression for a spherical wave in spherical coordinates

$$\psi(\mathbf{r}, t) = \frac{A_0}{r} \exp[ikr - i\omega t]$$

Note that $\phi = kr$ is not a scalar product of two vectors but a product of two scalar quantities. Again, $k = \omega/c = 2\pi/\lambda$, but $\nabla\phi = k\hat{r}$, where $\hat{r}$ is a unit vector in the radial direction. The inverse dependence of the amplitude on r is a result of energy conservation.

Example 8.6 Superposition of two plane waves

What happens if we superpose two plane waves?

$$\psi(\mathbf{r}, t) = A_0 \exp[i\mathbf{k}_1 \cdot \mathbf{r} - i\omega t] + A_0 \exp[i\mathbf{k}_2 \cdot \mathbf{r} - i\omega t].$$

Let us choose the wave vectors $\mathbf{k}_1$ and $\mathbf{k}_2$ to lie on the x–z plane with a common z component and equal but opposite x component (Fig. 8.10a):

$$\begin{aligned}\mathbf{k}_1 &= k\cos\theta\,\hat{z} + k\sin\theta\,\hat{x}\\ \mathbf{k}_2 &= k\cos\theta\,\hat{z} - k\sin\theta\,\hat{x}\end{aligned}$$

We have used the hat ˆ symbol (inverted caret) to indicate unit vectors. The resultant wave is then

$$\psi(\mathbf{r},t) = 2A_0\cos[k\sin\theta\,x]\exp[ik\cos\theta\,z - i\omega t]$$

The phase fronts are planes normal to the z axis, and the phase velocity is now $\omega/k\cos\theta = c/\cos\theta > c$. The phase velocity is greater than c because along z, the wavefront travels a distance $c/\cos\theta$ in one period (Fig. 8.10b).

8.3.2 Beam formation by superposition of plane waves

The last example of superposition of two plane waves serves well as an introduction to the subject of this section. Plane waves extend to all space, and are uniform in the direction transverse to the propagation direction, whereas an optical beam is confined in the transverse direction. However, as we have seen in Example 8.6, by superimposing two plane waves, we can obtain a resultant wave that varies sinusoidally in the transverse direction through interference of the two component plane waves. If we carry this one step further and superimpose many plane waves, it is possible, by interference, to construct any wave amplitude distribution in the transverse direction. The propagation of a confined wave is the core of the diffraction theory. A particular case is the Gaussian beam. For mathematical simplicity and ease of visualization, we will restrict ourselves to waves on the two dimensional x–z plane. Only in the final stage will we present the results for full three dimensional Gaussian beams.

Before going into detailed calculations, let us look at the last example again. By superimposing two plane waves, each of which propagates at an angle θ to the z axis, we obtain a resultant wave that propagates along the z axis with a phase velocity greater than c, and whose amplitude varies sinusoidally in the transverse x direction. A simple physical explanation can be given. The wave vectors k_1 and k_2 are drawn in Fig. 8.10a, together with their components in the x and z directions. Their common component in the z direction results in wave propagation in that direction. The equal and opposite components in the x direction form a standing wave that varies sinusoidally in that direction with a spatial frequency $k\sin\theta \simeq k\theta$ for small θ. We now come to a very important property of wave diffraction. Suppose, to confine the wave in the transverse direction, that we keep adding plane waves, each of which propagates at a different, small angle θ to the z axis, so that the amplitude adds constructively within the range $|x| < \Delta x$ and destructively outside. By the uncertainty principle which results from Fourier analysis and applies to this case as well, we have

$$\Delta(k\theta)\cdot\Delta x \gtrsim 1$$

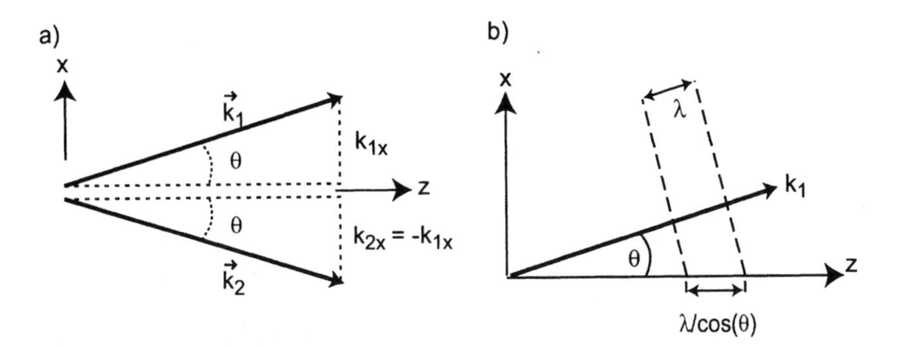

Figure 8.10: Superposition of two plane waves. (a) The wave vectors of two plane waves. They have equal components in the z direction; equal and opposite components in the x direction. The equal and opposite components in the x direction result in a standing wave. (b) The phase velocity along the z direction is greater than c, the speed of light; because in one period the wavefront of either component wave propagates a distance λ, but along the z axis, a distance of $\lambda/\cos\theta$.

or

$$\Delta\theta \gtrsim \frac{\lambda}{\pi 2\Delta x} \tag{8.8}$$

In words, confining a beam to a width of Δx requires a spread of plane waves over an angular range of at least $\lambda/2\pi\Delta x$. The angular spread means that the beam will eventually diverge with an angle $\Delta\theta$.

8.3.3 Fresnel integral and beam propagation: near field, far field, Rayleigh range

We now superimpose plane waves to form a beam. For mathematical simplicity, we look at beams only on the x–z plane. These beams are "sheets," infinite in extent in the y direction. The results for the more realistic waves in three dimensions are very similar and will be given afterwards. We choose the beam propagation direction to be z. Each component plane wave propagates at an angle θ to the z axis and has an amplitude $A(\theta)d\theta$, so that the resultant wave, with the harmonic time variation omitted, is

$$\psi(x,z) = \int d\theta A(\theta) \exp[ik\sin\theta\, x + ik\cos\theta\, z]$$

In what is called the *paraxial approximation*, $A(\theta)$ is significant only over a small angular range near zero; meaning, from Eq. 8.8, that the beam transverse dimension is large compared to the wavelength. The limits of integration can be extended to $\pm\infty$ for mathematical convenience if needed. We expand the trigonometrical functions to θ^2:

$$\begin{aligned}\psi(x,z) &\simeq \int d\theta A(\theta)\exp\left[ik\theta\, x + ik\left(1-\frac{\theta^2}{2}\right)z\right] \\ &= e^{ikz}\int d\theta A(\theta)\exp\left[ik\theta\, x - ik\frac{\theta^2}{2}\,z\right] \qquad (8.9)\end{aligned}$$

It is necessary to keep the quadratic term, otherwise the field would be independent of z other than the propagation factor e^{ikz}. The wave can be regarded as a plane wave e^{ikz} modulated by the integral in Eq. 8.9. Equation 8.9 completely describes the propagation of the wave if the wave is known at some point, say at $z = 0$. In fact, at $z = 0$, Eq. 8.9 is a Fourier transform

$$\begin{aligned}\psi(x,0) &\equiv \psi_0(x) \\ &= \int d\theta A(\theta)\exp[ik\theta x] \qquad (8.10)\end{aligned}$$

and we can find the angular distribution by inverse transformation:

$$A(\theta) = \frac{k}{2\pi}\int dx'\psi_0(x')\exp[-ikx'\theta]$$

Substitution of $A(\theta)$ back into Eq. 8.9 yields

$$\psi(x,z) = \frac{k}{2\pi}e^{ikz}\int d\theta \int dx'\psi_0(x')\exp\left[ik\theta x - ikx'\theta - ik\left(\frac{\theta^2}{2}\right)z\right] \qquad (8.11)$$

We can first integrate over θ and obtain the field at z as an integral over the field at $z = 0$, $\psi_0(x)$. The integral is performed by completing the square in the exponent, and is detailed in Appendix 8.C. The result is

$$\begin{aligned}\frac{k}{2\pi}e^{ikz}\int d\theta\exp\left[ik\theta x - ikx'\theta - ik\left(\frac{\theta^2}{2}\right)z\right] &= \sqrt{\frac{k}{i2\pi z}}\exp\left[ikz + \frac{ik(x-x')^2}{2z}\right] \\ &\equiv h(x-x',z) \qquad (8.12)\end{aligned}$$

which, is called the *impulse response*, or *kernel*, or *propagator*, or *Green's function*. It has a very simple physical interpretation—it is the field at (x, z) generated by a point source of unit strength at $(x', 0)$, and is a (two-dimensional) spherical wave in paraxial form. The field of a two-dimensional spherical wave (i.e., a circular wave) with its center at $(x', 0)$ is

$$\sqrt{\frac{1}{r}}\exp[ikr]$$

where $r = \sqrt{(x-x')^2 + z^2}$. (Instead of $1/r$ as in three dimensions, the amplitude decreases as $\sqrt{1/r}$ in two dimensions.) Near the z axis, $r \simeq z + (x-x')^2/(2z)$, and the spherical wave is approximately

$$\sqrt{\frac{1}{z}}\exp\left[ikz + \frac{ik(x-x')^2}{2z}\right]$$

the same expression as $h(x-x', z)$ in Eq. 8.12. Note that in expanding r, we have kept the quadratic term in the exponent, because that term, although small compared to z, may not be small compared to the wavelength so that when multiplied by k, it may not be a small number. Equation 8.9 becomes

$$\psi(x,z) = \int dx' \psi_0(x') h(x-x', z) \tag{8.13}$$

$$= \sqrt{\frac{k}{i2\pi z}} e^{ikz} \int dx' \psi_0(x') \exp\left[\frac{ik(x-x')^2}{2z}\right] \tag{8.14}$$

We will call this integral the *Fresnel integral*. It is the mathematical expression of *Huygens' principle*: the field at (x, z) is the sum of all the spherical waves centered at each previous point $(x', 0)$ whose strength is proportional to the field strength at $(x', 0)$.[1]

Equations 8.9 and 8.14 represent two equivalent ways to calculate wave propagation. Equation 8.9 calculates the wave from the angular distribution of its component plane waves. When the angular distribution is Hermite-Gaussian, a Gaussian beam results. Equation 8.14 calculates the wave field at a later point z from the field at the initial point $z = 0$. This is the traditional Fresnel diffraction theory. A Gaussian beam also results if ψ_0 is a Hermite-Gaussian.

Before any detailed calculations and examples, it is possible to get a good general idea on the propagation from these two formulations, and in the process, introduce the important concept of near and far fields. Let us first look at the wave in the near field, specifically, at a distance z small enough that the quadratic term in the exponent of Eq. 8.9 is much less than unity; then that term can be ignored, and

$$\psi(x,z) \simeq e^{ikz} \int d\theta A(\theta) \exp[ik\theta x]$$

$$= e^{ikz} \psi_0(x)$$

where the second line follows from Eq. 8.10. The near field, at zeroth approximation, is just the field at $z = 0$ multiplied by the propagation phase factor $\exp(ikz)$. We will examine the first order correction presently. Let us define "near" more precisely. It was determined by the condition

$$k\theta^2 z << 1$$

The question is then what is the maximum angle θ? It is not $\pi/2$; rather, it is the angular spread $\Delta\theta$ over which $A(\theta)$ is significantly different from zero. Put in another way, if the wave varies significantly within a distance Δx, then by Eq. 8.8, the inequality above becomes

$$\frac{\pi \Delta x^2}{\lambda} >> z/2$$

[1]The spherical wave is in paraxial form, and the interpretation for close z is rather subtle. See A. E. Siegman, *Lasers*, University Science Books, 1986 for full discussion on this point.

The quantity on the left, called the *Rayleigh range*, is the demarcation between near and far fields. A simple physical interpretation for this quantity is given below.

Let us look at the other limit, of large z or far field, using Eq. 8.14. When ψ_0 is confined to Δx and if z is large enough so that the quadratic factor is much less than unity

$$\frac{kx'^2}{(2z)} << 1$$

or

$$\frac{\pi \Delta x^2}{\lambda} << z$$

then it can be ignored and the integral becomes

$$\psi(x,z) \simeq \sqrt{\frac{k}{i2\pi z}} \exp\left[ikz + \frac{ikx^2}{2z}\right] \int dx' \psi_0(x') \exp\left(\frac{-ikxx'}{z}\right)$$

The integral is a Fourier transform. The *magnitude* of the far field is the magnitude of the Fourier transform of the field at $z = 0$. It is not an exact Fourier transform because of the quadratic phase factor $kx^2/(2z)$ in front of the integral above.[2]

Let us return to the near field and calculate the first-order correction. For small $z = \Delta z$, we can expand the exponent in Eq. 8.9:

$$\begin{aligned}\psi(x, \Delta z) &\simeq e^{ik\Delta z} \int d\theta A(\theta) \left(1 - ik\frac{\theta^2}{2}\Delta z\right) \exp\left[ik\theta x\right] \\ &= e^{ik\Delta z}\psi_0(x) - e^{ik\Delta z}\frac{ik\Delta z}{2} \int d\theta A(\theta)\theta^2 \exp\left[ik\theta x\right]\end{aligned}$$

The integral in the last line is

$$\int d\theta A(\theta)\theta^2 \exp[ik\theta x] = -\frac{1}{k^2}\frac{\partial^2}{\partial x^2} \int d\theta A(\theta) \exp[ik\theta x] = -\frac{1}{k^2}\frac{\partial^2 \psi_0(x)}{\partial x^2}$$

so that the correction term is

$$\frac{i\Delta z}{2k}\frac{\partial^2 \psi(x,0)}{\partial x^2}$$

Note that it is in quadrature with the zeroth-order term, if ψ_0 is real. The second derivative can be viewed as a diffusion operator, as the second derivative of a bell-shaped function is negative in the center and positive in the wings so that when added to the original function, the center is reduced whereas the wings are increased. The quadrature means that the correction is in phase, not in magnitude. This phase diffusion is the cause of diffraction.

[2] In fact one can perform an exact Fourier transformation by a lens of focal length f to correct the quadratic phase factor. See Problem 8.5.

We can generalize a little further. Near $z = 0$, from above, we obtain

$$\psi(x, \Delta z) \simeq e^{ik\Delta z}\left[\psi(x,0) + \frac{i\Delta z}{2k}\frac{\partial^2\psi(x,0)}{\partial x^2}\right]$$

Suppose that we write ψ as a plane wave e^{ikz} modulated by a slowly varying function $u(x, z)$

$$\psi(x, z) \equiv u(x, z)e^{ikz}$$

then

$$u(x, \Delta z) - u(x, 0) = \frac{i\Delta z}{2k}\frac{\partial^2 u(x,0)}{\partial x^2}$$

This relationship was derived at one particular point on the z axis: $z = 0$. However, there was no particular requirement for this point, and the relationship applies at any z. So taking the limit $\Delta z \to 0$, we have

$$2ik\frac{\partial u}{\partial z} + \frac{\partial^2 u}{\partial x^2} = 0 \tag{8.15}$$

This equation is called the *paraxial wave equation.* It is an approximate form of the scalar wave equation, and has the same form as the Schrödinger equation for a free particle. The equation can be generalized to three dimensions by a similar derivation:

$$2ik\frac{\partial u}{\partial z} + \left(\frac{\partial^2 u}{\partial x^2} + \frac{\partial^2 u}{\partial y^2}\right) = 0 \tag{8.16}$$

The Fresnel integral is the solution of the paraxial wave equation with the given boundary condition of ψ at $z = 0$. In Appendix 8.A, we outline how a three-dimensional wave can be built up from two-dimensional waves. The resulting Fresnel integral in three dimensions is

$$\psi(x, y, z) = \tag{8.17}$$

$$\frac{k}{i2\pi z}e^{ikz}\int dx'dy'\psi_o(x', y')\exp\left[\frac{ik(x-x')^2}{2z}\right]\exp\left[\frac{ik(y-y')^2}{2z}\right]$$

where $\psi_0\,(x, y)$ is the field distribution at $z = 0$. Note that, as required by energy conservation, the field, in three dimensions, decays as $1/z$, not as $\sqrt{1/z}$ in two dimensions. And note that the three-dimensional impulse response is essentially the product of two 2D impulse responses. Finally, we discuss the physical meaning of the Rayleigh range. The reference of near or far field is always to a particular plane, chosen to be $z = 0$ in this case. At $z = 0$, the field extends to Δx. Consider the two points $(0, 0)$ and $(\Delta x, 0)$ and the distances from them to a point $(0, z)$ on axis (Fig. 8.11). The distances are z and $\sqrt{\Delta x^2 + z^2} \simeq z + \Delta x^2/(2z)$, respectively. If a spherical wave emanates from $(0, 0)$ and another from $(\Delta x, 0)$, when the waves reach $(x = 0,\ z)$, they will have picked up, from propagation, phase factors of kz and $kz + k\Delta x^2/(2z)$, respectively. The difference between these phase factors is $k\Delta x^2/(2z)$. This difference is unity at the Rayleigh range. The phase difference is negligible in the far field but significant in the near field.

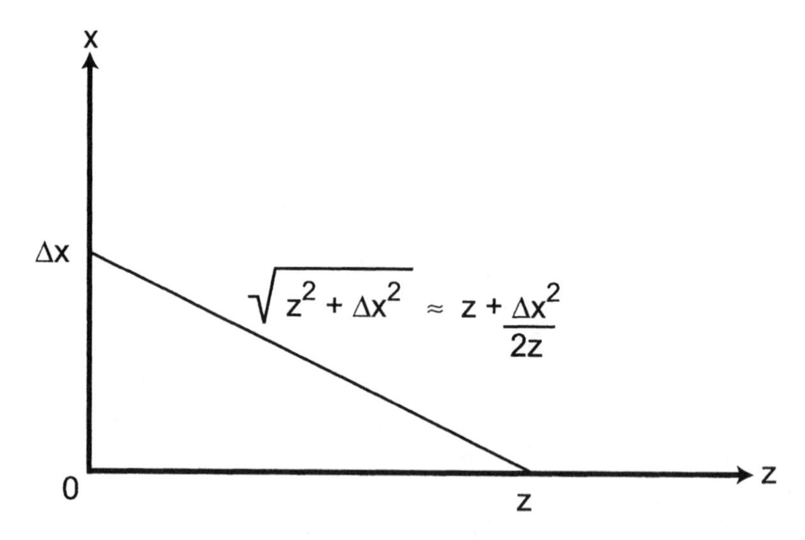

Figure 8.11: Rayleigh range. The distance from $(0, 0)$ to $(0, z)$ is z. The distance from $(0, \Delta x)$ to $(0, z)$ is approximately $z+\Delta x^2/(2z)$. The difference is $\Delta x^2/(2z)$. Hence a wave originating from $(0, 0)$ and another from $(0, \Delta x)$ will acquire a phase difference of $k\Delta x^2/(2z)$ when they reach $(0, z)$. The phase difference is unity when z equals the Rayleigh range.

8.3.4 Applications of Fresnel diffraction theory

We will apply the Fresnel diffraction integral, Eq. 8.14, to a few cases, to illustrate its use and the difference between wave and geometric optics.

Diffraction through a slit

Suppose that a uniform plane wave of amplitude A traveling in the z direction impinges on a screen at $z = 0$, which is opaque except for an opening at $|x| < \Delta x/2$. What is the field at $z > 0$?

The exact field distribution at the opening $z = 0$ is a difficult boundary-value problem. However, intuition suggests that it is equal to the incident amplitude A for $|x| < \Delta x$ and zero otherwise. The field at $z > 0$ is, by the Fresnel integral (Eq. 8.14), with $\psi_0(x) = A$

$$\psi(x, z) = \sqrt{\frac{k}{i2\pi z}} e^{ikz} A \int_{-\Delta x}^{\Delta x} dx' \exp\left[\frac{ik(x - x')^2}{2z}\right]$$

The integral cannot be evaluated in closed form. For numerical integration, it is convenient to normalize the spatial variables. A natural scale for the transverse coordinate is the aperture size Δx. Denoting $x/\Delta x$ and $x'/\Delta x$ by $\overline{x}$ and $\overline{x}'$, the field becomes

$$\psi(x, z) = e^{ikz} A \sqrt{\frac{1}{i\pi\overline{z}}} \int_{-1}^{1} d\overline{x}' \exp\left[\frac{i(\overline{x} - \overline{x}')^2}{\overline{z}}\right]$$

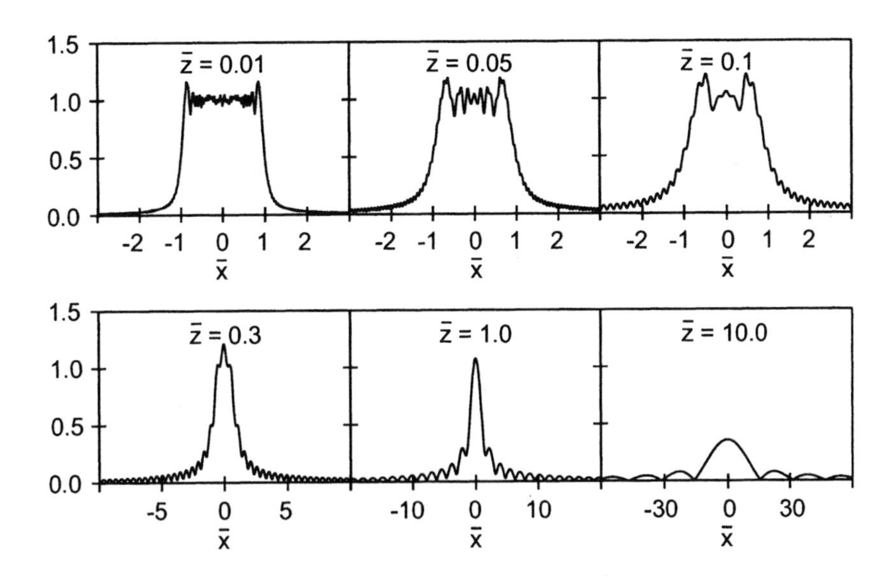

Figure 8.12: Diffraction through a slit: field magnitude at different distances from the slit. Vertical axis: normalized magnitude of the transmitted wave through a slit at different distances from the slit. The distance is normalized to the Rayleigh range. Horizontal axis: transverse distance normalized to half of the slit width. Note the change of scale.

where the normalized axial coordinate $\overline{z}$ is z/b and b is the Rayleigh range

$$b = \frac{\pi \Delta x^2}{\lambda}$$

For large $\overline{z}$ or in the far field, we ignore the quadratic term $\overline{x}'^2$, and the integral becomes a Fourier transform, as discussed in Section 8.3.3. For small $\overline{z}$, the exponent is large and the exponential function fluctuates rapidly, except when $\overline{x}$ is close to $\overline{x}'$, that is, the integrand approximates a delta function, and we are in the near field where the wave changes little. When λ approaches zero compared to the aperture size, the Rayleigh range approaches infinity, and the fringes of the far field, which are characteristics of waves, will not appear. The magnitude of ψ is plotted for several $\overline{z}$ in Fig. 8.12. Note how the field evolves from its initial box distribution to the final sinc function; the change is most rapid where the field changes fastest, that is, at the edges.

In terms of geometric optics, rays pass through the slit undeflected, so that the intensity distribution faithfully duplicates the slit. This distribution is a fair approximation for the near field (short wavelength limit) but fails completely in the far field.

Problem 8.3 *Consider the problem of diffraction through a slit.*

(a) Consider the diffraction through a slit, 0 $< x <$ d, with d$\rightarrow \infty$. The problem then becomes one of diffraction through a straight edge. Evaluate and plot the transmitted field versus x at a few points from the edge. Compare the diffracted field from a straight edge relate it to that of the near field diffracted from a slit.

(b) A wave is incident normal to a perfectly conducting screen that has an opening at -d/2 $< x <$ d/2. Calculate the reflected wave. How is the reflected wave related to the transmitted wave?

(c) A wave is incident normally on a strip of perfect conductor at -d/2$<$ x $<$d/2. Calculate the reflected wave and transmitted wave. How are these waves related to those in (b)?

The pinhole camera

The pinhole camera is a rudimentary imaging device. There is an opening (pinhole) on one face of an otherwise closed box, and the image is projected by the object through the pinhole onto the inside face opposing the opening. The pinhole camera is used in X-ray imaging. The vision of many organisms is afforded by pin-hole cameras.[3] To illustrate the effects of geometric and wave optics, we will calculate the size of the pinhole for maximum spatial resolution. Let the radius of the pin-hole be a and the distance between the hole and the image plane d (Fig. 8.13a). Light from the object at a far distance D ($>> d$) may be considered as emanating from many points on the object. A point illuminating the opening will, by geometric optics when a is large, create an image of dimension approximately equal to the pinhole size a (Fig. 8.13b). Better resolution, that is, smaller image size of the point, is achieved by reducing the hole. When a is reduced sufficiently, however, according to wave optics, light after passing through the pinhole will diverge, with an angle of about $\lambda/(\pi a)$; therefore the size of the image is about $d \cdot \lambda/(\pi a)$, which increases with decreasing hole size (Fig. 8.13c). The minimum image size is then obtained when the geometric optical image size is equal to the wave optical image size, $a \simeq d \cdot \lambda/(\pi a)$, or $d \simeq \pi a^2/\lambda$, the Rayleigh range of the pinhole (Fig. 8.13d). The same conclusion can be reached using the Fresnel integral, but there is no closed-form solution when the opening is "hard." When the transmission through the opening is approximated by a Gaussian, a closed form solution can be found and will be treated with Gaussian beamsas discussed below.

Action of a thin lens

A lens is "thin" if its thickness is much smaller than the Rayleigh range as defined by of aperture of the lens, so that in passing through the lens, a beam is not diffracted. The beam, however, picks up a phase factor that varies quadratically in the transverse dimension. That phase factor depends on the focal length of the lens and affects the subsequent diffraction of the beam. As an example, take

[3]See the fascinating account "The Forty-fold Path to Enlightenment" in Richard Dawkins, *Climbing Mount Improbable*, Norton, 1996

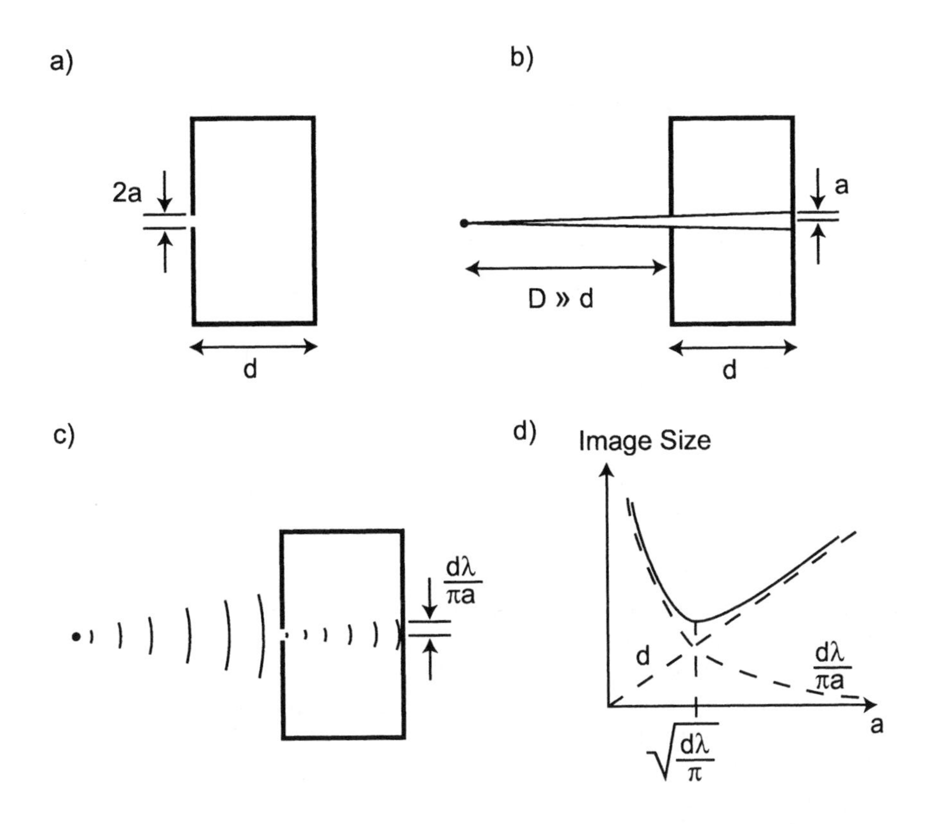

Figure 8.13: Pinhole camera and the image of a distant point. (a) The pinhole camera. (b) The image of a distant point according to geometric optics. (c) The image of a distant point by diffraction through the hole. (d) Image size of a distance point versus hole radius. The image size is minimum at a radius such that the depth of the camera is equal to the Rayleigh range of the hole.

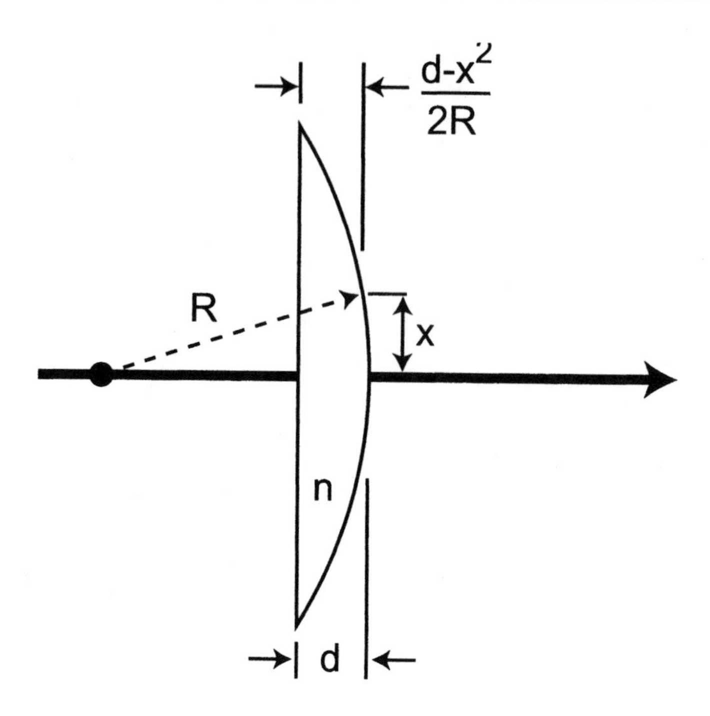

Figure 8.14: The plano-convex lens. One surface of the lens is plane, the other spherical with radius R. The thickness of the lens at the center is d. At a distance x from the center, the thickness is $d - x^2/2R$

the case of a planoconvex lens (Fig. 8.14). It is a dielectric material one surface of which is plane, and the other spherical of radius R. The index of refraction is n and the thickness at $x = 0$ is d. The difference between d and the thickness of the dielectric at x is $R - \sqrt{R^2 - x^2} \simeq x^2/(2R)$. Define the entrance and exit planes of the lens as the plane of the flat surface and the plane touching the apex of the lens, respectively; then a plane wave in the z-direction passing through the lens picks up at x a phase factor of $kx^2/(2R)$ from traveling through air, and $nk[d - x^2/(2R)]$ from traveling in the dielectric. The total phase shift is $nkd - (n-1)kx^2/(2R)$. Plane waves traveling at small angles to z pick up the same phase factor to order x^2. The same consideration applies to lens of other quadratic surfaces. The constant phase factor nkd can be ignored, and the proportionality constant of the quadratic phase shift defines the focal length f of the lens, so that the exit field from and the entrance field to a lens are related by

$$\psi_{\text{exit}} = \exp\left[-i\frac{kx^2}{2f}\right]\psi_{\text{entrance}} \tag{8.18}$$

As a simple check, a plane wave incident on the lens emerges as a spherical wave with the center at a distance f from the lens.

Problem 8.4 *Consider the thin lens and diffraction.*

(a) A handheld magnifier, with a focal length of 5 cm and a diameter of 4 cm, is used to focus sunlight which on a bright day can be as intense as 100 mW/cm^2. What is the light intensity at focus? What is the depth of focus? Use an average wavelength of 0.5 μm.

(b) The published angular resolution of the Hubble Space Telescope is 0.053 arcsec for visible light. How does that compare with the angular resolution limited by diffraction?

Imaging with a thin lens—the lens law revisited

We will look again at the lens law, Eq. 8.4, derived earlier by ray optics. From the wave theory, a couple of important points will emerge that cannot be obtained or are not obvious from ray optics. One is the resolution limit imposed by the size of the lens. The other is the nature of the image field, which turns out not to be exactly a scaled replica of the object field even if the lens is ideal. Again, the object is at a distance d_o from the lens. The field of the object is $\psi_0(x)$. Right in front of the lens, the field, according to Eq. 8.13, is

$$\psi_1(x) = \int dx' \psi_0(x') h(x - x', d_o)$$

After the lens, the field is, according to Eq. 8.18,

$$\psi_1(x) \exp\left[-i\frac{kx^2}{2f}\right].$$

At the image distance d_i from the lens, the field, by Eq. 8.13 again, becomes,

$$\begin{aligned} \psi(x) &= \int dx'' \psi_1(x'') \exp\left[-i\frac{kx''^2}{2f}\right] h(x - x'', d_\mathrm{i}) \\ &= \int dx' dx'' \psi_0(x') \exp\left[-i\frac{kx''^2}{2f}\right] h(x - x'', d_i) h(x'' - x', d_o) \end{aligned}$$

In each of the impulse response functions h there is a quadratic phase factor; there is another quadratic phase factor from the lens. Together, the quadratic phase factors add up to

$$ik\left[-\frac{x''^2}{2f} + \frac{(x - x'')^2}{2d_\mathrm{i}} + \frac{(x'' - x')^2}{2d_\mathrm{o}}\right]$$

The quadratic terms in x'' will cancel and the integration over the linear phase terms x'' can be performed if the lens law holds, that is, if

$$-\frac{1}{f} + \frac{1}{d_\mathrm{i}} + \frac{1}{d_\mathrm{o}} = 0$$

The integration over x'' is

$$\int dx'' \exp ik\left\{-\frac{xx''}{d_i} - \frac{x'x''}{d_o}\right\} \tag{8.19}$$

The integration limits are the lens aperture. But suppose for the moment the lens is large enough to approximate the limits by infinity, then the integral yields a delta function

$$\delta\left(\frac{x}{d_i}+\frac{x'}{d_o}\right)$$

and the image field is, apart from a intensity scaling factor and a constant phase factor $k(d_\mathrm{o}+d_\mathrm{i})$

$$\begin{aligned}\int dx'\psi_0(x')\exp\left[ik\left\{\frac{x^2}{2d_\mathrm{i}}+\frac{x'^2}{2d_\mathrm{o}}\right\}\right]\cdot\delta\left(\frac{x}{d_\mathrm{i}}+\frac{x'}{d_\mathrm{o}}\right) \\ = \exp\left[ik\left\{\frac{x^2}{2d_\mathrm{i}}\left(\frac{d_\mathrm{o}}{f}\right)\right\}\right]\cdot\psi_0\left(-\frac{d_\mathrm{o}x}{d_\mathrm{i}}\right)\end{aligned}$$

Thus the image field is a scaled version of the object field, multiplied by a quadratic phase factor. This quadratic phase factor is usually—but not always—inconsequential. Indeed, this additional phase factor is crucial in one convenient way of analyzing multimirror optical resonators, discussed below.

Let us return to the finite lens aperture and Eq. 8.19. If the lens diameter is D, then the integral becomes, instead of a delta function, a sinc function:

$$\frac{\sin\left[\frac{kD}{2}\left(\frac{x}{d_\mathrm{i}}+\frac{x'}{d_\mathrm{o}}\right)\right]}{\frac{kD}{2}\left(\frac{x}{d_\mathrm{i}}+\frac{x'}{d_\mathrm{o}}\right)}$$

and the image field is

$$\int dx'\psi_0(x')\exp\left[ik\left\{\frac{x^2}{2d_\mathrm{i}}+\frac{x'^2}{2d_\mathrm{o}}\right\}\right]\cdot\mathrm{sinc}\left[\frac{kD}{2}\left(\frac{x}{d_\mathrm{i}}+\frac{x'}{d_\mathrm{o}}\right)\right]$$

which has the form of a convolution of the object field with the sinc function. The sinc function integrates $\psi_0(x')$ over its (the sinc function's) width $d_\mathrm{o}\lambda/D$. Two points in ψ_0, separated by $\Delta x'$, are separated in the image only if they are separated by more than $d_\mathrm{o}\lambda/D$, or

$$\frac{\Delta x'}{d_\mathrm{o}}>\frac{\lambda}{D}$$

The left-hand side is the angle subtended by the two points in the object at the lens, and the right-hand side, the diffraction angle of the lens aperture. The resolution improves with larger aperture or shorter wavelength.

Problem 8.5 *The Fresnel integral and the Fourier transform.*

Prove, by direct application of the Fresnel integral, that the field with a focal length behind a lens is the Fourier transform of that field with a focal length in front of the lens. The proof is rather formal. For a physical discussion, see H. A. Haus, Waves and Fields in Optoelectronics, Prentice-Hall, 1984.

Diffraction through a thin periodic structure

A widely used optical component is the grating. The grating can take many forms, but basically it is a periodic structure that, through multiple interference, sends an incident wave to different directions depending on the wavelength of the wave. Most practical gratings have periods on the order of the wavelengths for which they are designed, so that wave diffracted from the grating can spread over a large angle, and the Fresnel diffraction theory developed may not apply. However, as seen immediately below, the diffracted wave consists of well separated, narrow lobes; over each lobe, Fresnel diffraction is valid. Moreover, the essential features of gratings come from their periodicity, and we will consider diffraction through a thin structure whose period Λ is larger than the optical wavelength λ. Let a plane wave $\exp(ikz)$ be incident on a thin periodic structure located at $z = 0$ whose transmission is

$$T(x) = T(x + n\Lambda)$$

where n is an integer ranging from 1 to N, and N is the number of periods through which the incident wave passes. After emerging from the structure, the plane wave is modulated in the transverse direction, and is

$$\psi_0(x) = T(x)$$

According to Fresnel diffraction theory, when the wave propagates further in the z direction, it will remain approximately unchanged for a distance $\pi\Lambda^2/\lambda$, the Rayleigh range corresponding to the structure period Λ. At a greater distance in the Fraunhoffer region, where $z >> \pi(N\Lambda)^2/\lambda$, the Fresnel integral simplifies and the wave becomes

$$\begin{aligned}
\psi(x, z) &\simeq \sqrt{\frac{k}{i2\pi z}} \exp\left[ikx + i\frac{kx^2}{2z}\right] \int dx' \psi_0(x') \exp\left(-i\frac{kxx'}{z}\right) \\
&\propto \int dx' \psi_0(x') \exp\left(-i\frac{kxx'}{z}\right) \\
&= \sum_{n=1}^{N} \int_{(n-1)\Lambda}^{n\Lambda} dx' T(x') \exp\left(-i\frac{kxx'}{z}\right)
\end{aligned}$$

where we have broken the integral over the whole periodic structure into N sections, each over one period. The integral can be recast by making use of the periodicity of T:

$$\begin{aligned}
&\int_{(n-1)\Lambda}^{n\Lambda} dx' T(x') \exp\left(-i\frac{kxx'}{z}\right) \\
&= \int_0^{\Lambda} dx' T\left(x' - (n-1)\Lambda\right) \exp\left[-i\frac{kxx'}{z}\right] \exp\left[-i\frac{kx(n-1)\Lambda}{z}\right] \\
&= \left[\int_0^{\Lambda} dx' T(x') \exp\left(-i\frac{kxx'}{z}\right)\right] \exp\left[-i\frac{kx(n-1)\Lambda}{z}\right]
\end{aligned}$$

The integral within the square brackets is the Fourier transform of one period of T. Thus the wave emerging from one period is a product of two terms; one is the Fourier transform of one period, which is the same irrespective of the position of the period, and the other is a phase factor that depends on the position of the period but independent of the specific transmission function. Putting the expression back into the sum, we have

$$\begin{aligned}\psi(x,z) &\propto \left[\int_0^{\Lambda} dx' T(x') \exp\left(-i\frac{kxx'}{z}\right)\right] \sum_{n=1}^{N} \exp\left[-i\frac{kx(n-1)\Lambda}{z}\right] \\ &= \left[\int_0^{\Lambda} dx' T(x') \exp\left(-i\frac{kxx'}{z}\right)\right] \left[\frac{\sin\left[Nk\Lambda x/2z\right]}{\sin[k\Lambda x/2z]}\right]\end{aligned}$$

Thus the far field is a product of two terms. The first term is the Fourier transform of one period. The second, called the *form factor*, is an interference term that is the result solely of the periodicity and independent of the details of the structure. It is the second term, when the structure is used as a grating, that is responsible for its ability to resolve wavelengths. The spatial dependence of the field can be written in terms of the angle $\theta = x/z$. As will be seen immediately below, the form factor is negligibly small except near several angles, called *orders*, at which all the waves from all the periods add in phase. The first term in θ varies much more slowly than the second, as the former is the far field of just one period whereas the latter is that of N periods. Over the angular range where the form factor varies rapidly from almost zero to a maximum and back to zero again, the first term hardly changes. Thus the first term, or the specific transmission on one period, only determines the efficiency of the diffraction into the orders.

Let us now examine the form factor and see how it determines the wavelength resolving power of the structure. In terms of $\theta = x/z$ and $\lambda = 2\pi/k$, it is

$$F(\lambda,\theta) \equiv \frac{\sin\left[N\pi\Lambda\theta/\lambda\right]}{\sin\left[\pi\Lambda\theta/\lambda\right]}$$

Where $F(\lambda, 0) = N$ for all λ. This is called the *zeroth order.* It rapidly drops to zero at a small angle $\delta\theta = (\lambda/\Lambda)/N$. As θ increases further, F remains small until we come to the first order, at $\theta = \lambda/\Lambda$ when its magnitude becomes N again. Now if two nearly equal wavelengths λ_1 and λ_2 pass through the structure, they will be diffracted into two different angles $\theta_i = \lambda_i/\Lambda$ in the first order, and they can be separated only if $|\theta_1 - \theta_2| = |\lambda_1 - \lambda_2|/\Lambda \geq \delta\theta = (\lambda/\Lambda)/N$, or only if

$$\frac{|\lambda_1 - \lambda_2|}{\lambda_1} \geq \frac{1}{N}$$

Thus the resolution of the structure is equal to the number of periods that the incident wave passes through. This result can also be derived from physical considerations. At the first-order angle, waves diffracted from each period of the grating add in phase. Consider two waves from the bottom of two consecutive

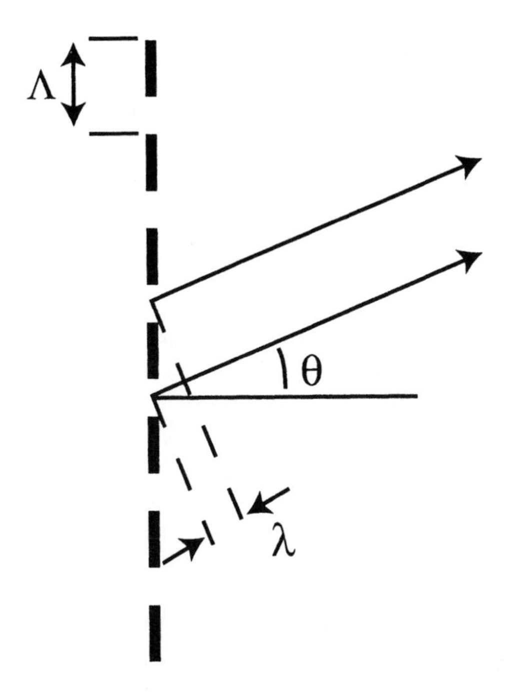

Figure 8.15: Diffraction from a grating. The grating period is Λ. At a diffraction order, the waves diffracted from all the periods add in phase. When two waves originate at points one period apart, their path difference must be whole wavelengths. For first order, the path difference is one wavelength λ. The diffraction angle θ is therefore given by $\lambda = \Lambda\,\theta$.

periods (Fig. 8.15). The path difference must be λ, that is, $\Lambda\theta \simeq \lambda$. On the other hand, the diffracted beam diverges with an angle $\theta_d \simeq \lambda/(N\Lambda)$. A different wavelength λ', to be separated from λ, must be diffracted to angle θ' so that $|\theta' - \theta| > \theta_d$, or $|\lambda' - \lambda|/\Lambda > \lambda/(N\Lambda)$, the same result as above.

Problem 8.6 *Diffraction and polarization:*

So far, we have not considered the effect of polarization in diffraction. Suppose a grating is made of parallel long metallic strips. Will it better diffract waves polarized parallel or perpendicular to the strips?

Problem 8.7 *Plane waves and lenses:*

(a) A plane wave is incident on a lens (of focal length f) at an angle θ from the optical axis. Prove that it is focused at a point $f\theta$ above the axis on the focal plane (Fig. 8.16).

(b) A field distribution $\psi_0(x)$ can be decomposed as a superposition of plane waves at different angles from the optical axis. If a lens is placed at some distance from it, use the result in (a) to show that the field on the focal plane is the far field of ψ_0.

(c) A beam with irregular profile can be smoothed and recollimated with a pair of lenses and a pinhole. The positive lenses are separated by the sum of their focal lengths, with the pinhole between the lens at the focal plane of the first lens. Explain the smoothing.

8.3.5 Further comments on near and far fields, and diffraction angles

In many cases, the field $\psi_0(x)$ changes significantly over a distance much smaller than the extent of $\psi_0(x)$. The grating just discussed above is an example: a plane wave passing through it varies significantly in x within one period Λ, which is much smaller than the size of the grating, $N\Lambda$. Now, $\psi_0(x)$ has two characteristic lengths, Λ and $N\Lambda$. Which one determines the near and far fields? For the near field, the wave transmitted by each period does not change appreciably within the Rayleigh range as determined by Λ. Farther than that, the waves from adjacent periods begin to overlap and interfere, and we are no longer in the near field. But we are not yet in the far field. The far field, it will be recalled, is given by the condition that the quadratic phase factor $kx'^2/2z$ in the Fresnel integral be much less than unity, so $k(N\Lambda)^2/2z << 1$, or $z >> \pi(N\Lambda)^2/\lambda$. The far field is determined by the grating size $N\Lambda$; the near field, by one period Λ.

The far field of the diffracted light from a grating may be too far for the laboratory. For example, for 1 μm wavelength light covering 1 cm of grating, the far field is ~300 m from the grating. The standard method to observe the far field in a shorter distance is to use a lens. The field distribution on the focal plane (one focal length behind the lens) is the far field. The proof follows from the fact that a plane wave traveling at an angle θ from the optical axis is focused by the lens to a point $x = f\theta$ on the focal plane (Fig. 8.16) where f is

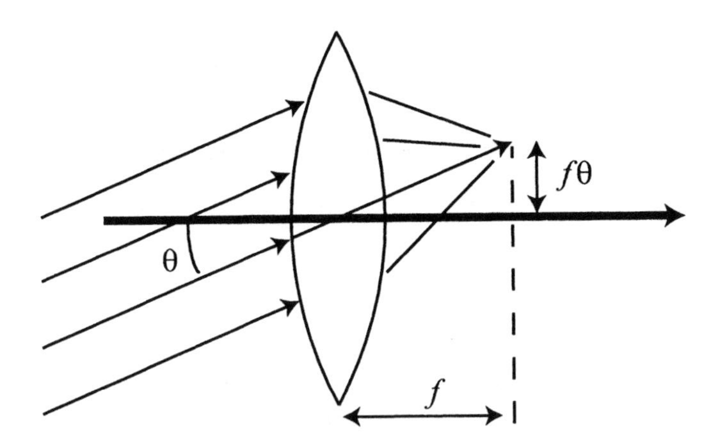

Figure 8.16: Focusing of an oblique plane wave by a lens. A plane wave traveling at an angle θ from the optical axis is focused by a lens (of focal length f) to a point on the focal plane, at a distance $f\,\theta$ from the axis.

the focal length of the lens (see Problem 8.7a). Now if we consider that wave leaving the grating, $\psi_0(x)$, as a superposition of plane waves at different angles with distribution $A(\theta)$, then the intensity at the focal plane represents $|A(\theta)|^2$. From Eq. 8.10, $A(\theta)$ is the Fourier transform of $\psi_0(x)$, as is the far-field of $\psi_0(x)$ (within a phase factor).

Another interesting property of grating is the far-field diffraction angles. The far field is, from above, proportional to

$$\left[\int_0^{\Lambda} dx' T(x') \exp(-ik\theta x')\right] \left[\frac{\sin[Nk\Lambda\theta/2]}{\sin[k\Lambda\theta/2]}\right]$$

where $\theta = x/z$. The second factor is periodic. In each period there is a narrow lobe with an angular width $\lambda/(N\Lambda)$, which is the diffraction angle of the whole grating. The first factor is the Fourier transform of one period, so that the angular width is $\sim \lambda/\Lambda$, much wider than the lobes of the second factor. The diffraction pattern is therefore a series of lobes in different angles (orders) whose widths are determined by the whole grating, with an envelope whose width is determined by one period.

8.4 The Gaussian Beam

A very important class of waves, the Gaussian beam, is obtained when the angular distribution $A(\theta)$ is a Hermite-Gaussian function. The Gaussian beam is what comes out of almost all lasers, because it is the normal mode of optical resonators formed by spherical mirrors. We will examine the two-dimensional fundamental Hermite-Gaussian beam in detail, then give the expressions for the three-dimensional Gaussian beams. We will see that inside a resonator consisting of two spherical mirrors, a Gaussian beam can reproduce itself in one round trip if the optical frequency is correct, that is, a mode is formed.

8.4.1 The fundamental Gaussian beam in two dimensions

When the angular distribution is a Gaussian,

$$A(\theta) = A_0 \exp\left[-\left(\frac{\theta}{\theta_0}\right)^2\right],$$

where A_0 and $\theta_0 << 1$ are constants, we can extend the limits of the integration over θ to $\pm\infty$, since $A(\theta)$ is significant over a small-angle range of θ_0, and the integration can be carried out in closed form by completing the square in the exponent to obtain the full expression for $\psi(x, z)$. In anticipation of this rather complex result, let us first introduce one of the several parameters by Fourier-transforming $A(\theta)$ to obtain the field at $z = 0$ (Eq. 8.10):

$$\begin{aligned}\psi_0(x) &= \int d\theta A_0 \exp\left[-\left(\frac{\theta}{\theta_0}\right)^2 + ik\theta x\right] \\ &= \sqrt{\pi}\theta_0 A_0 \exp\left[-\left(\frac{k\theta_0}{2}x\right)^2\right]\end{aligned}$$

The field at $z = 0$ is a Gaussian, with width w_0 which will be called the *beam waist*

$$w_0 \equiv \frac{2}{k\theta_0} = \frac{\lambda}{\pi\theta_0}$$

Note that

$$\theta_0 = \frac{\lambda}{\pi w_0} \tag{8.20}$$

is the manifestation of the uncertainty relationship for this particular beam. ψ_0 can be written in terms of w_0,

$$\psi_0(x) = A_0 \frac{\lambda}{\sqrt{\pi} w_0} \exp\left[-\left(\frac{x}{w_0}\right)^2\right]$$

Related to w_0 is the Rayleigh range b,

$$b \equiv \frac{\pi w_0^2}{\lambda} = \frac{w_0}{\theta_0}$$

The field at $z \neq 0$ can be obtained from either Eq. 8.9 or 8.14. The result is

$$\begin{aligned}\psi_0(x, z) = \frac{\sqrt{2\pi} A_0}{\sqrt[4]{k^2(z^2 + b^2)}} \exp\left[-\left(\frac{x}{w(z)}\right)^2\right] \times \\ \exp\left[ikz + i\frac{kx^2}{2R(z)}\right] \exp\left[-i\frac{1}{2}\arctan\left(\frac{z}{b}\right)\right]\end{aligned} \tag{8.21}$$

where we added a subscript 0 to $\psi(x,z)$ in anticipation of higher-order beams. The field ψ_0 has been separated into four factors: (1) amplitude, (2) beam size, (3) radius of curvature, and (4) phase factor.

The first factor

$$\frac{\sqrt{2\pi}A_0}{\sqrt[4]{k^2(z^2+b^2)}} \tag{8.22}$$

is a direct result of energy conservation. As the beam expands, the field amplitude must decrease to keep the total power constant. This becomes apparent once it is recast in terms of the spot size defined immediately below.

The second factor

$$\exp\left[-\left(\frac{x}{w(z)}\right)^2\right]$$

quantifies the transverse extent of the wave as a function of z. The spot size $w(z)$ is

$$w(z) = \sqrt{w_0^2+(z\theta_0)^2} \tag{8.23}$$

$$= w_0\sqrt{1+\left(\frac{z}{b}\right)^2} \tag{8.24}$$

Note that at large distance $z \gg b$, $w(z) \to z\,\theta_0$, and the spot size diverges linearly with an angle θ_0.

The third factor defines the spherical phase front, approximately quadratic near the z axis:

$$\exp\left[ikz+i\frac{kx^2}{2R(z)}\right]$$

The radius of curvature is

$$R(z) = z+\frac{b^2}{z} \tag{8.25}$$

At large z, $R(z) \to z$, and the large spherical wavefront centers at the origin.

The fourth factor

$$\exp\left[-i\frac{1}{2}\arctan\left(\frac{z}{b}\right)\right] \tag{8.26}$$

is a phase shift that results in a change of phase velocity on the optical axis.

The first, energy conservation factor can be rewritten as

$$\frac{\lambda A_0}{\sqrt{\pi}w_0}\sqrt{\frac{w_0}{w(z)}}. \tag{8.27}$$

Note that the parameters that vary along the z axis, $w(z)$, $R(z)$, and $\arctan(z/b)$ all have the Rayleigh range b as the characteristic length. They are plotted in Fig. 8.17. We define the Gaussian modulation function $u_0(x,z)$ as

$$u_0(x,z) \equiv \sqrt{\frac{w_0}{w(z)}}\cdot\exp\left[-\left(\frac{x}{w(z)}\right)^2+i\frac{kx^2}{2R(z)}-i\frac{1}{2}\arctan\left(\frac{z}{b}\right)\right] \tag{8.28}$$

so that the Gaussian beam is a plane wave $\exp[ikz]$ modulated by u_0.

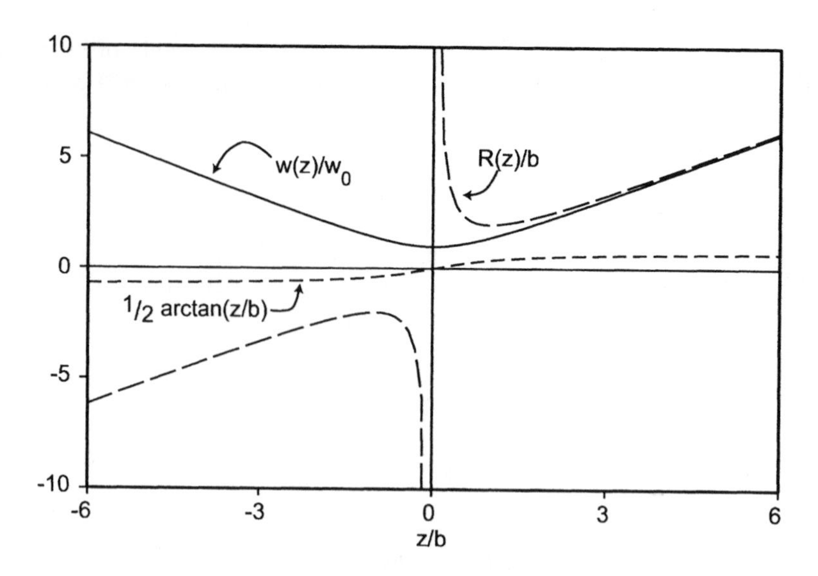

Figure 8.17: Parameters of a Gaussian beam. Parameters of a two-dimensional Gaussian beam as a function of distance from the beam waist. The distance is normalized to the Rayleigh range b. (1) Beam size $w(z)$ normalized to the beam waist w_o; (2) radius of curvature $R(z)$ normalized to b; (3) axial phase shift $\frac{1}{2}\arctan(z/b)$ in radians. The asymptotic values at large $|z/b|$ is $\pm\pi/4$.

Problem 8.8 *The Gaussian beam and a Fabry–Perot interferometer.*

A Fabry–Perot interferometer consists of two parallel, partially transmitting plane mirrors. A Gaussian beam is incident onto the interferometer with the beam waist on the front mirror. In the far field of the transmitted beam, fringes (rings of bright and dark rings) appear. Explain. What is the spacing between two adjacent bright rings?

8.4.2 Higher-order Gaussian beams in two dimensions

Beside the fundamental Gaussian beam, there are infinitely many other confined beams, among which of particular importance are the higher-order Gaussian beams. The fundamental and higher-order Gaussian beams form an orthogonal and complete set; any other beam can be expressed as a superposition of these beams, and they are the eigenmodes or normal modes of optical resonators formed of spherical mirrors that are used almost universally in lasers.

The higher-order Gaussian beams $\psi_m(x,z)$ are fundamental Gaussian beams amplitude-modulated by Hermite polynomials H_m and phase-modulated by an additional phase factor $-m \cdot \arctan(z/b)$:

$$\begin{aligned}\psi_m(x,z) &= H_m\left(\frac{\sqrt{2}x}{w(z)}\right)\exp\left[-im\arctan\left(\frac{z}{b}\right)\right]\cdot\psi_0(x,z)\\ &\equiv u_m(x,z)\cdot\exp(ikz) \qquad (8.29)\end{aligned}$$

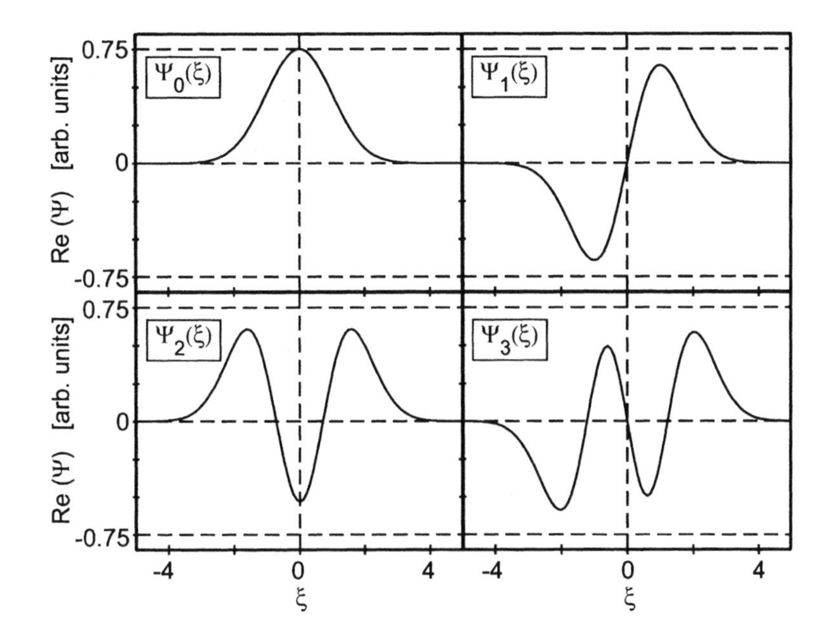

Figure 8.18: The first four Hermite-Gaussians.

where we have defined the Hermite-Gaussian beam modulation function $u_m(x, z)$

$$u_m(x, z) \equiv \sqrt{\frac{w_0}{w(z)}} H_m\left(\frac{\sqrt{2}x}{w(z)}\right) \exp\left[-\left(\frac{x}{w(z)}\right)^2\right] \cdot \exp\left[i\frac{kx^2}{2R(z)}\right] \cdot \exp[-i\phi_m(z)] \tag{8.30}$$

and the axial phase

$$\phi_m(z) \equiv \left(m + \frac{1}{2}\right) \arctan\left(\frac{z}{b}\right)$$

These beams are obtained from Hermite-Gaussian angular distributions. The Hermite-Gaussians are the same functions seen in solutions of the quantum-mechanical harmonic well problem. The first few Gaussian beams are plotted in Fig. 8.18.

8.4.3 Three-dimensional Gaussian beams

With initial Hermite-Gaussian distributions in Eq. 8.18, three-dimensional Gaussian beams are formed that, in Cartesian coordinates, are products of two 2D Gaussian beams:

$$\psi_{mn}(x, y, z) = u_m(x, z) \cdot u_n(y, z) \cdot \exp(-ikz) \tag{8.31}$$

In this expression, it is possible to have different beam waists w_{0x} and w_{0y} in the x and $y$directions, and therefore different Rayleigh ranges b_x and b_y, in $u_m(x, z)$ and $u_n(y, z)$, respectively. The asymmetry results in elliptical spot sizes and wavefronts. The phase angle when $m = n = 0$ is $\arctan(z/b)$, twice that in two dimensions. The fundamental, symmetric Gaussian beam is

$$\psi_{00} = \frac{w_0}{w(z)} \exp\left[-\frac{x^2+y^2}{w(z)^2} + i\frac{k(x^2+y^2)}{2R(z)} - i \arctan\left(\frac{z}{b}\right)\right]$$

In cylindrical coordinates, different expressions of Gaussian beams emerge, but they can be regarded as superpositions of Cartesian Gaussian beams. For example, the "donut mode" is a superposition of two Cartesian beams of equal beam waist:

$$\begin{aligned}\psi_{10}(x,y,z) + i\psi_{01}(x,y,z) &\sim (x+iy)\exp\left[-\frac{(x^2+y^2)}{w^2(z)}\right] \\ &\sim r\exp\left[-\frac{r^2}{w^2(z)}\right].\end{aligned}$$

8.4.4 Gaussian beams and Fresnel diffraction

Since the Gaussian beams are special cases of Fresnel diffraction, and they form a complete set, diffraction of an initial, arbitrary field distribution $\Psi(x,y)$ can be handled using Gaussian beams as well as the Fresnel integral, by expansion in a series of Gaussian beams:

$$\Psi(x,y,z=0) = \sum A_{mn}\psi_{mn}(x,y,z=0)$$

In subsequent diffraction, the field becomes

$$\Psi(x,y,z) = \sum A_{mn}\psi_{mn}(x,y,z)$$

Since the transverse field distribution of the Gaussian beams remain Hermite Gaussians in propagation, the diffraction of $\Psi(x,y,z)$ is a result *solely* of the relative axial phase shifts $(m+n)\arctan(z/b)$. To illustrate this important point, let us look at the passage of a plane wave through a slit again. Let us keep only the first two nonzero terms in the series for $\Psi(x, z=0)$ and just make the expansion coefficients equal:

$$\begin{aligned}\Psi(x,0) &= \psi_0(x,z=0) + \psi_2(x,z=0) \\ &= \exp\left[-\frac{x^2}{w_0^2}\right]\left[1 + H_2\left(\frac{\sqrt{2}x}{w_0}\right)\right]\end{aligned}$$

The magnitude of Ψ is plotted in Fig. 8.19. At $z > 0$, the axial phase of each Hermite-Gaussian ψ_m is different, so

$$\begin{aligned}\Psi(x,z) &= \psi_0(x,z) + \psi_2(x,z) \\ &= \sqrt{\frac{w_0}{w}}\exp\left[-\frac{x^2}{w^2} + i\frac{kx^2}{2R} - i\frac{1}{2}\arctan\left(\frac{z}{b}\right)\right] \times \\ &\quad \left[1 + H_2\left(\frac{\sqrt{2}x}{w}\right)\exp\left[-i2\arctan\left(\frac{z}{b}\right)\right]\right]\end{aligned}$$

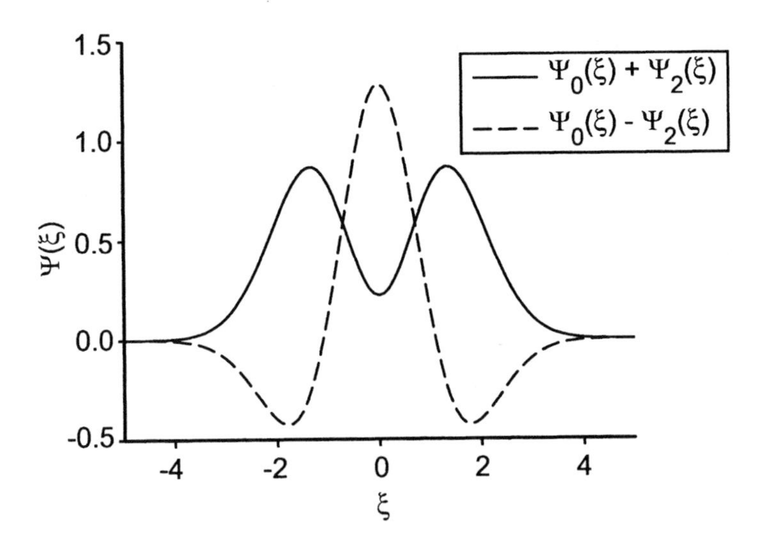

Figure 8.19: Diffraction through a slit as a superposition of two Hermite-Gaussian beams. The wave immediately after passing through the slit is approximated by adding the fundamental and second order Hermite-Gaussian beams (solid line). In the far field, the two Gaussian beams acquire a relative phase shift of π and they subtract (broken line). The wave now resembles a sinc function.

In the far field $z >> b$, $\arctan(z/b) \rightarrow \pi/2$, and the sign of ψ_2 changes:

$$\Psi(x, z >> b) \propto \exp\left[-\frac{x^2}{w^2}\right]\left[1 - H_2\left(\frac{\sqrt{2}x}{w}\right)\right]$$

The magnitude of Ψ is again plotted, and we can see the shape of the expected far field pattern begins to emerge.

8.4.5 Beams of vector fields, and power flow

So far, we have treated the optical field as a scalar quantity. In an area small compared with the spot size, the field is approximately linearly polarized, and we can take ψ to be the electric field amplitude. In building up two-dimensional beams from plane waves, one can superimpose vector, instead of scalar, plane waves, and the resultant fields will be vectorial. Suppose the electric field of a component plane wave with its wave vector inclined at angle θ to the z axis lies on the x–z plane. The previous scalar amplitude $A(\theta)$ is now interpreted as the magnitude of the electric field. The z component of the electric field is $-A(\theta)\sin(\theta) \simeq -\theta A(\theta)$. The x component of the electric field is $A(\theta)\cos(\theta) \simeq$

$A(\theta)$. The total electric field component in the z direction, E_z, is

$$\begin{aligned} E_z(x,z) &\simeq -\int d\theta A(\theta)\theta \exp[ik\sin(\theta)x + ik\cos(\theta)z] \\ &\simeq -\frac{1}{ik}\frac{\partial}{\partial x}\int d\theta A(\theta)\exp[ik\theta x + ik\cos(\theta)z] \\ &= -\frac{1}{ik}\frac{\partial\psi(x,z)}{\partial x} \end{aligned}$$

Similarly, the total electric field component in the x direction is

$$E_x(x,z) \simeq \psi(x,z)$$

The magnetic field in this case is entirely in the y direction and proportional to ψ.

It turns out that the expressions above for the electric field components are valid also in three dimensions, if we take $\psi(x,y,z)$ to be the x component of the vector potential $\mathbf{A}$. The magnetic field can be calculated from

$$\begin{aligned} \mathbf{B}(x,y,z) &= \nabla \times \mathbf{A} \\ &\simeq ikA_x\hat{\mathbf{y}} - \frac{\partial A_x}{\partial y}\hat{\mathbf{z}} \end{aligned}$$

and the electric field can be calculated from the magnetic field through one of Maxwell's equations:

$$\begin{aligned} -\frac{i\omega}{c}\mathbf{E} &= \nabla \times \mathbf{B} \\ &\simeq k^2 A_x\hat{\mathbf{x}} + ik\frac{\partial A_x}{\partial x}\hat{\mathbf{z}} \end{aligned}$$

Near the optical axis, the beam is linearly polarized. The average intensity flow is given by the Poynting vector:

$$\frac{1}{2}\mathbf{E}\times\mathbf{B}^* \simeq \frac{1}{2}k^2 A_x A_x^*\hat{\mathbf{z}} - \frac{1}{2}ikA_x^*\frac{\partial A_x}{\partial x}\hat{\mathbf{x}} + \frac{1}{2}ik\frac{\partial A_x}{\partial y}A_x^*\hat{\mathbf{y}}$$

The power flow is dominant in the propagation direction, along which the Poynting vector is real, indicating real power flow. In the transverse directions, it is complex; imaginary power flow indicates oscillatory power storage, not power flow or dissipation, and is a signature of evanescent waves. The complexity of power flow is important in guided waves. In unbound structures it has become important in light of recent developments in near-field optical microscopy where the traditional practical limit of resolution to one wavelength has been circumvented. The source is of dimensions much smaller than a wavelength. An object close to the source converts the evanescent fields to propagating fields which are detected and analyzed at a distant point.

8.4.6 Transmission of a Gaussian beam

We saw in Section 8.2.1 that the transmission of rays through optical elements is facilitated, in the paraxial approximation, by the use of the $ABCD$ matrices. It is remarkable that the same $ABCD$ matrices can be applied to the Gaussian beam, although the parameters related by the matrices are different. The parameter that the $ABCD$ matrix transforms is the q parameter. To introduce the parameter, let us derive the fundamental Gaussian beam from a spherical wave with a mathematical transformation. The spherical wave near the z axis is

$$\frac{1}{z}\exp\left(ikz + i\frac{kr^2}{2z}\right)$$

We translate the origin by an *imaginary* length

$$z \longrightarrow z - ib \equiv q \tag{8.32}$$

where b is a constant. The paraxial spherical wave is transformed into the fundamental Gaussian beam. Comparing the quadratic factor in the exponent

$$i\frac{kr^2}{2(z-ib)} \equiv i\frac{kr^2}{2q} = i\frac{kr^2 z}{2(z^2+b^2)} - \frac{kr^2 b}{2(z^2+b^2)}$$

with the quadratic factors in the exponent of the Gaussian beam modulation function, Eq. 8.28, it can be seen that

$$\frac{1}{q} = \frac{1}{R(z)} - i\frac{\lambda}{\pi w^2(z)} \tag{8.33}$$

At $z = 0, q = -ib$, $R{=}\infty$ and $w(0) = w_o$, so $b = \pi w_o^2/\lambda$, the Rayleigh range. The imaginary part can be regarded as the inverse of the Rayleigh range of the Gaussian beam at z.

If the q parameter of a Gaussian beam before and after an optical element is q_1 and q_2, respectively, and the optical element is characterized in ray optics by the matrix elements $ABCD$, then the q parameter is transformed by

$$q_2 = \frac{Aq_1 + B}{Cq_1 + C} \tag{8.34}$$

The proof of this general relationship is beyond the scope of this text and can be found in the references. The reader can, using ray optics, prove that the relationship holds for spherical waves where $q = R$, the radius of the wave, and thus obtain a plausibility argument by viewing the inverse q-parameter as a generalized inverse radius of curvature. We only need two special cases, both of which can be verified directly: two points in space separated by a distance d, and a thin lens of focal length f. Cascading of optical elements are handled by multiplication of the matrices of the respective optical elements, as in ray optics.

In free space, $q_2 = q(z+d)$, $q_1 = q(z)$. By Eq. 8.32, $q_2 = q_1 + d$. For free space, the $ABCD$ matrix elements are $A = 1$, $B = d$, $C = 1$, $D = 0$. The same result is obtained from Eq. 8.34. In going through a thin lens, the change in spot size w is negligible. The radius of curvature is changed from $1/R$ to $1/R - 1/f$. So $1/q_2 = 1/q_1 - 1/f$. The $ABCD$ matrix elements for a lens are $A = 1$, $B = 0$, $C = -1/f$, $D = 1$; therefore Eq. 8.34 yields $q_2 = q_1/(-q_1/f + 1)$ or $1/q_2 = 1/q_1 - 1/f$. Proof of cascading $ABCD$ matrices for a series of optical elements is left, again, to the references.

Finally, one must remember that the third important parameter of a Gaussian beam, the axial phase shift $m \arctan(z/b)$ is not included in the q parameter. If we are only dealing with a single Gaussian beam, an axial phase shift is usually not an issue. However, the phase shift must be tracked carefully when the beam propagates in a resonator, or if the beam is a superposition of Gaussian beams because the diffraction of the beam is determined *completely* by the relative axial phase shifts of the component Gaussian beams.

8.4.7 Mode matching with a thin lens

A common operation in the laboratory is matching the beam from a laser onto an optical component like a fiber or a resonator. First consider mode-matching a laser beam to a single mode fiber. By that we mean we want to couple as much light into the fiber as possible. When a wave propagates inside, and is guided by, a single mode fiber, the optical field is well defined. It can be approximated by

$$E_{\text{fib}}(x, y)e^{i\beta z}$$

where, near the center of the fiber, $E_{\text{fib}}(x, y)$ is approximately a Gaussian $\exp[-(x^2 + y^2)/a^2]$ of waist a. Now, if we shine the beam directly onto the fiber, chances are that more of the light will be reflected and scattered than going into the fiber and propagating as a mode with the field distribution above. In mathematical terms, the incident field is being decomposed as a superposition of the fields of the modes of the fiber, including those that radiate away. To maximize the coupling, the field amplitude $E_{\text{in}}(x, y)$ of the laser beam right at the entrance of the fiber must be as close to E_{fib} as possible. This is measured quantitatively by the *coupling coefficient*, which is proportional to

$$\int dx\, dy E_{\text{in}}(x, y) E^*_{\text{fib}}(x, y)$$

(The reader may note the similarity in finding Fourier expansion coefficients.) Since $E_{\text{in}}(x, y)$ is a Gaussian beam, it might be thought that if its spot size were made equal to the spot size a of the fiber field, then the coupling would be maximized. This alone is not enough. As can be verified by performing the integration above, the radius of curvature must also be infinite, that is, the beam waist must be at the entrance to the fiber. So the coupling problem reduces to transforming a Gaussian beam of waist w_0 to another value a. This can often be accomplished with a single lens. The solution according to ray optics is one of imaging by the lens, with w_0/a equal to d_1/d_2, where d_1 is the distance between the lens and the waist w_0 and d_2 that between the lens and the waist

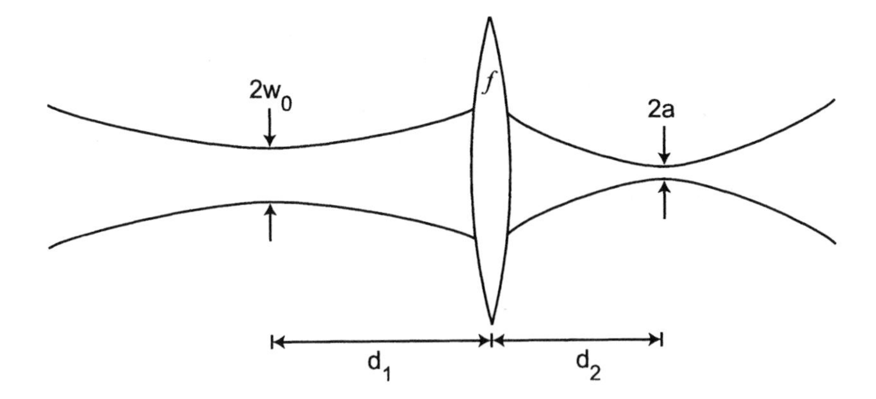

Figure 8.20: Modematching by transformation of beam waist with a lens. A Gaussian beam of waist w_o is transformed by a lens to a Gaussian beam of waist a. With the focal length of the lens as a parameter, the distances d_1 and d_2 can be found using the $ABCD$ matrices. The fiber is positioned so that the transformed beam enters it with waist a.

a. We saw in the previous section on Fresnel diffraction theory that the image has an additional spherical wavefront, and this solution is therefore inaccurate. The accurate solution by Gaussian beam optics can be obtained using $ABCD$ matrices. First, by measuring the spot size at two places, the waist w_0 and its position can be deduced. Then set $1/q_1$ to $-i\lambda/(\pi w_0^2)$ or $q_1 = ib_1$. At a distance d_1 from the beam waist there is a lens of focal length f. At a distance d_2 from the lens the beam is focused again and the beam waist is a and

$$q_2 = i\frac{\pi a^2}{\lambda} \equiv ib_2$$

(see Fig. 8.20). Through the distance d_1, the lens, and the distance d_2, the sequence of matrices is

$$\begin{pmatrix} 1 & d_2 \\ 0 & 1 \end{pmatrix} \cdot \begin{pmatrix} 1 & 0 \\ -\frac{1}{f} & 1 \end{pmatrix} \cdot \begin{pmatrix} 1 & d_1 \\ 0 & 1 \end{pmatrix} =$$

$$\begin{pmatrix} 1 - \frac{d_2}{f} & d_1 + d_2 - \frac{d_1 d_2}{f} \\ -\frac{1}{f} & 1 - \frac{d_1}{f} \end{pmatrix} \equiv \begin{pmatrix} A & B \\ C & D \end{pmatrix} \tag{8.35}$$

q_2 is related to q_1 through the $ABCD$ matrix by Eq. 8.34. The real and imaginary parts of the equation yield two equations relating the three quantities f, d_1 and d_2. This means that the solution is not unique. In practice, the choice of f is limited by available lenses, so let us find d_1 and d_2 using f as a parameter, subject to the constraint that d_1 and d_2 be positive. First, because q_2 is purely imaginary, we have

$$\mathrm{Re}(q_2) = \mathrm{Re}\left(\frac{Aq_1 + B}{Cq_1 + D}\right) = \mathrm{Re}\left(\frac{iAb_1 + B}{iCb_1 + D}\right) = 0$$

which yields

$$\frac{d_2}{f} = \frac{\left(\frac{b_1}{f}\right)^2 + \frac{d_1}{f}\left(\frac{d_1}{f} - 1\right)}{\left(\frac{b_1}{f}\right)^2 + \left(\frac{d_1}{f} - 1\right)^2} \tag{8.36}$$

Substituting d_2 into the imaginary part of q_2 yields

$$\frac{b_2}{b_1} = \left(\frac{a}{w_0}\right)^2 = \frac{1}{\left(\frac{b_1}{f}\right)^2 + \left(\frac{d_1}{f} - 1\right)^2} \tag{8.37}$$

from which d_1/f can be found, and substituting it back in Eq. 8.36 from which d_2/f can be found. The results are

$$d_1 - f = \pm\frac{\omega_0}{a}\sqrt{f^2 - b_1 b_2}$$

$$d_2 - f = \pm\frac{a}{\omega_0}\sqrt{f^2 - b_1 b_2}$$

Thus the focal length of the lens must be greater than $\sqrt{b_1 b_2}$. Ostensibly there are two solutions. When ω_0 and a are vastly different, however, only the positive-sign solutions are valid. In this case, one of d_1 and d_2 is very large, the other very close to f. If the distance is too large to be practical, a second lens may be introduced. A limiting case is $f^2 >> b_1 b_2$, then the solutions lead to $d_2/d_1 \simeq a/w_0$, as predicted by ray optics using the lens equation.

The solutions for d_1 and d_2 can also be applied to a system consisting of more than one lens, like the compound lens discussed in Example 8.2, if f is taken as the effective focal length of the system, and d_1 and d_2 are measured from the principal planes.

The solutions can also be applied to mode matching with a resonator. In its simplest form, the resonator consists of a pair of mirrors between which a beam self-reproduces as it bounces back and forth. As seen below, the beam between the mirrors is a Gaussian beam if the mirrors are spherical. The beam waist is either inside the resonator, or at a point outside that can be found by extending the Gaussian beam beyond the mirrors. Mode matching to the resonator is accomplished by transforming the original Gaussian beam so that the new waist coincides with the resonator mode beam waist in both position and size.

Problem 8.9 *Modematching and the asymmetric Gaussian.*

The field distribution at the output of a semiconductor laser can be approximated by an asymmetric Gaussian

$$\exp\left[-\left(\frac{x}{a}\right)^2\right]\exp\left[-\left(\frac{y}{b}\right)^2\right]$$

where $a > b$. The output beam propagates in the z direction.

(a) Calculate and plot the spot size on the x–z plane and on the y–z plane as a function of z, with $\lambda = a/2 = b/10 = 1.5$ *μm.*

(b) Mode-match the beam to a fiber with a spherical lens and a cylindrical lens. A cylindrical lens can be formed by splitting a cylindrical rod lengthwise. The core of the fiber is 90 μm in diameter.

8.4.8 Imaging of a Gaussian beam with a thin lens

Let us now discuss briefly the image of a Gaussian beam. Since a Gaussian beam maintains its intensity profile of a Hermite-Gaussian throughout its propagation distance, and passing the beam through a lens does not change the intensity profile, the meaning of imaging is at first sight not at all clear. The meaning becomes clear when we take not one, pure Gaussian beam but a superposition of Gaussian beams, that is, if we expand the object field in a superposition of Gaussian beams of different orders. As mentioned earlier, the axial phase of each order propagates differently. Imaging then means regrouping these axial phases to their original values of zero, modulus π (not 2π, as the image may be inverted). Using the lens law and the $ABCD$ matrices, it is an exercise in algebra to prove the last statement.

8.4.9 The pinhole camera revisited

The problem of the resolution of the pinhole camera can also be solved with a Gaussian beam. If we approximate the field at the pinhole by a Gaussian with a waist a equal to the half of the diameter of the hole, then the subsequent propagation is that of a Gaussian beam. In particular, the spot size at the back wall, after traveling the depth d of the camera, $w(d)$, is given by

$$w^2(d) = a^2 \left(1 + \frac{d^2}{b^2}\right)$$

where $b = \pi a^2/\lambda$ is the Rayleigh range of the hole. We can view the first term on the right-hand side, which increases with a, as the contributing term from geometric optics, and view the second term, which decreases with a, from diffraction. Minimizing $w^2(d)$ with respect to a^2 leads to $d = b$, the same result as obtained before.

8.5 Optical Resonators and Gaussian Beams

A resonator serves several important, related functions. It allows energy storage; it discriminates signals; it provides positive feedback. The resonator is one of two indispensable elements of an oscillator.[4] Light is on resonance with the resonator if it reproduces itself on one round trip, with a possibly diminished amplitude due to loss.[5] The reproduction has two conditions: (1) the spatial

[4]The other indispensable element is gain.

[5]It cannot reproduce itself in more than one round trip if it cannot reproduce itself in one.

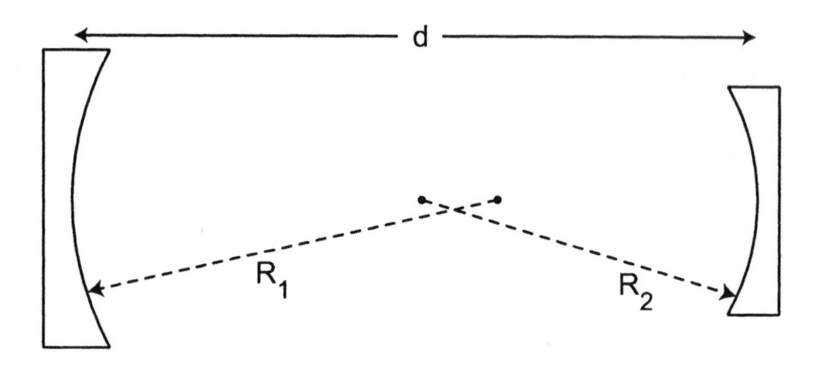

Figure 8.21: The two-mirror resonator: two mirrors separated by a distance d. The mirrors have intensity reflectivities R_1 and R_2. The two parameters d/R_1 and d/R_2 determine the modes of the resonator.

distribution of the field must reproduce itself, and (2) the total optical phase in one round trip must be an integral number of 2π. Each possible self-reproducing field distribution is called a *mode* of the resonator. For a given mode, only light of particular frequencies, called resonance frequencies, satisfies the condition on the total optical phase.

The simplest optical resonator consists of a pair of mirrors aligned on a common axis facing each other. Light traveling along the axis is reflected back from one mirror to the other. The resonator is completely characterized by the distance between the mirrors d, the radii of curvature of the mirrors, R_1 and R_2, and any additional loss suffered by the light in one round trip. The parameters d/R_1 and d/R_2 determine whether light can reproduce itself in a round trip between the mirrors. The condition of self-reproduction is called the *stability condition.* We will discuss the two-mirror resonator in detail. Multimirror resonators can be analyzed by reducing them to equivalent two-mirror resonators using imaging, as discussed below.

Several important parameters characterize a resonator. The *quality factor* Q and the related *finesse* $\mathcal{F}$ quantify the frequency selectivity of the resonator, or equivalently, the storage time of the resonator once an amount of energy is injected inside. The quality factor Q is equal to the number of cycles the optical field oscillates, and $\mathcal{F}$ the number of round trips the optical field makes, before the energy is depleted. The frequency-selective characteristics of the optical resonator is periodic in frequency. The periodicity in frequency is called the *free spectral range* (FSR). The approximate number of modes supported by the resonator in one FSR is given by the *Fresnel number* N_f. These parameters will be discussed in due course.

8.5.1 The two-mirror resonator

A resonator consists of two spherical mirrors facing each other (Fig. 8.21). A light beam starting at a location inside toward one mirror is reflected to the other and back to the original location. When the beam reproduces itself in one round trip, it is a mode of the resonator. Not all beams do; it turns out that each of the Gaussian beams, fundamental and high-order, does, and they are called the normal modes of the resonator. The parameters of the resonator, mirror radii R_1 and R_2, and mirror separation d, define the parameters of the Gaussian beams, the beam waist, and the location of the beam waist. These parameters are determined by the boundary conditions at the mirrors.

Boundary condition at the mirror

The requirement of self-reproduction leads to the boundary condition at the mirror that immediately before and immediately after the mirror the field be identical, except for a possible constant phase shift caused by a dielectric mirror, for example. Consider a Gaussian beam incident on a mirror of radius R. Recall that a mirror of radius R acts like a lens of focal length $R/2$, and a light passing through a lens picks up a curvature $1/f$. Right before the mirror, let the radius of curvature be R'. On reflection from the mirror, the radius of curvature is changed to R'' given by

$$-\frac{1}{R''} = \frac{1}{R'} - \frac{2}{R}$$

The negative sign on the left-hand side is due to the fact that the beam has changed its propagation direction on reflection. The requirement of self-reproduction means $R'' = R'$; therefore the equation above leads to

$$R' = R$$

In other words, at the mirror of a resonator, the radius of curvature of a mode must be equal to the radius of curvature of the mirror.

Stability conditions of a resonator

Consider a resonator with mirrors of radii R_1 and R_2 separated by a distance d. Let the beam waist of a Gaussian mode be at a distance d_1 from the mirror R_1 (Fig. 8.22). From the boundary conditions at the mirrors, we obtain

$$\begin{aligned} -R_1 &= -d_1 - \frac{b^2}{d_1} \\ R_2 &= (d - d_1) + \frac{b^2}{(d - d_1)^2} \end{aligned}$$

where $b = \pi w_0^2/\lambda$ is the Rayleigh range of the Gaussian mode. If the beam waist is located outside the resonator, say, to the left, then d_1 is negative. The two unknowns defining the mode, b and d_1, can be found from these two equations. Eliminating b^2 from the equations yields

$$d_1 = d\frac{R_2 - d}{R_1 + R_2 - 2d}$$

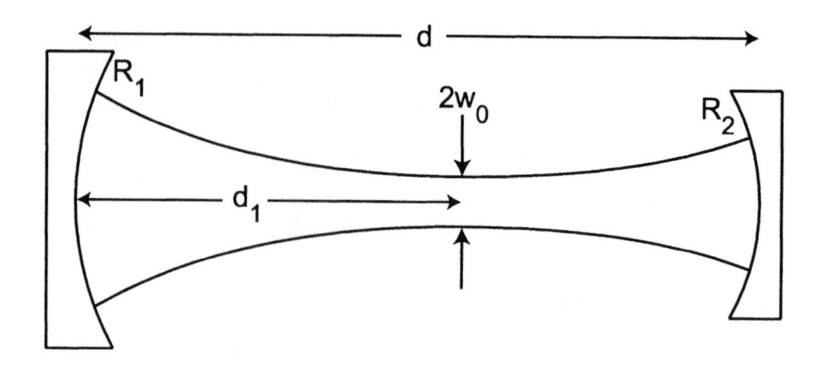

Figure 8.22: Gaussian beam inside a resonator: the beam waist w_0 and its location d_1 are determined by the resonator parameters d/R_1 and d/R_2.

Substituting d_1 into the first equation yields

$$b^2 = d\frac{(R_1 - d)(R_2 - d)(R_1 + R_2 - d)}{(R_1 + R_2 - 2d)^2}$$

That b must be real and positive leads to the resonator stability condition. Although three parameters define the resonator, the stability condition depends only on the relative magnitude of the mirror radii to the mirror separation, and it is more convenient to rewrite b in terms of two parameters g_1 and g_2, defined as

$$g_1 \equiv 1 - d/R_1, \; g_2 \equiv 1 - d/R_2 \tag{8.38}$$

After some algebraic manipulation, the formula for b^2 becomes

$$b^2 = d^2 \frac{g_1 g_2 (1 - g_1 g_2)}{(g_1 + g_2 - 2g_1 g_2)^2} \tag{8.39}$$

Thus the stability condition is

$$0 \leq g_1 g_2 \leq 1 \tag{8.40}$$

Figure 8.23 depicts the regions of stability where this relationship is satisfied on the g_1–g_2 plane. The resonator configuration in each of the stable region is illustrated in Fig. 8.24.

Unstable resonators

There are special resonators that fall outside the stability regions, called *unstable resonators*. They are used in pulsed, high-powered lasers. The reader is referred to Professor Siegman's book, (*Lasers*, University Science Books, 1986).

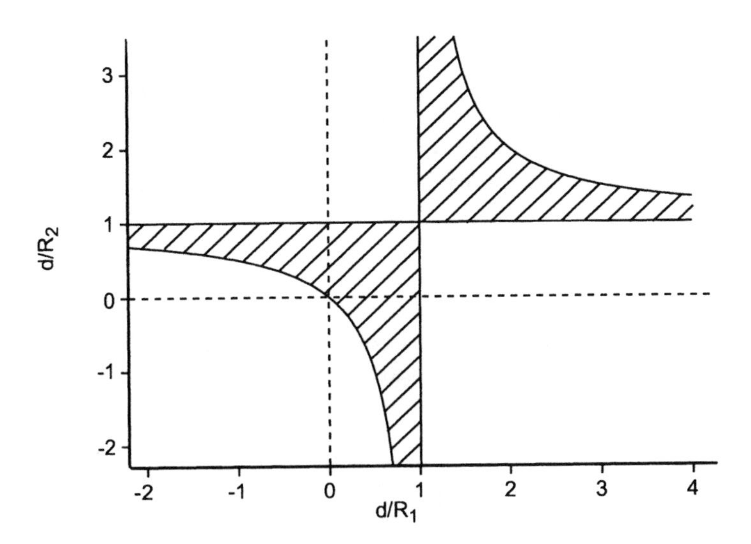

Figure 8.23: Stability diagram of a two-mirror resonator: resonators with parameters shown in the shaded regions are stable.

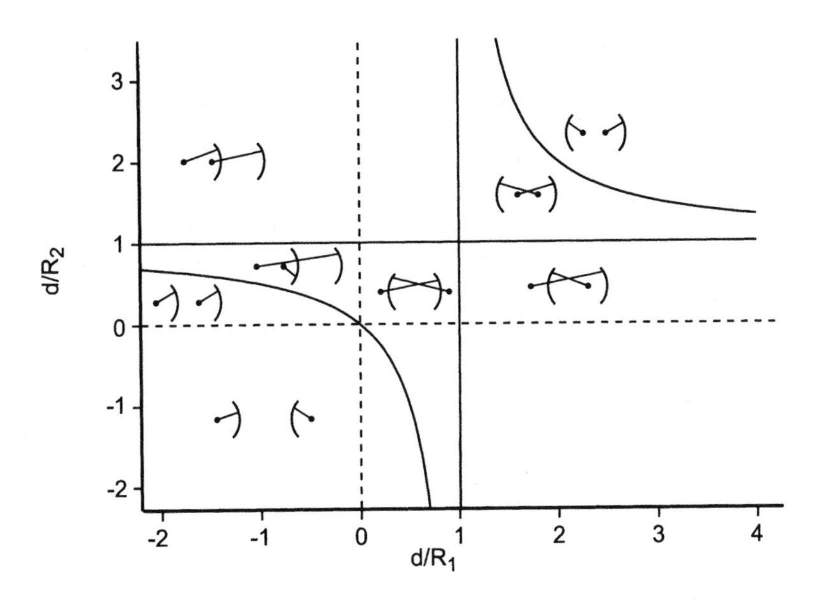

Figure 8.24: Resonator configurations: resonator configurations in different parts of the stability diagram: the centers of mirror curvature are shown with dots.

Resonance frequencies, the confocal resonator, and the scanning Fabry-Perot

Self-reproduction of the optical field includes the axial phase factor, which must change by an integral number of 2π after one round trip. This condition leads to the resonance frequencies of the resonator modes. The total axial phase shift of the (mn)-*th* order Gaussian beam is

$$kz + \phi_{mn}(z) = kz - (m+n+1)\arctan\left(\frac{z}{b}\right) \tag{8.41}$$

Since the different higher-order ($m \neq 0$ or $n \neq 0$) modes have different phase shifts, they have different resonance frequencies. As the axial coordinate is measured from the beam waist, the total axial phase shift in one round trip, is, from Eq. 8.41,

$$\begin{aligned} & 2kd - 2\phi_{mn}(d_1) - 2\phi_{mn}(d-d_1) \\ = \; & 2kd - 2(m+n+1)\left[\arctan\left(\frac{d_1}{b}\right) + \arctan\left(\frac{d-d_1}{b}\right)\right] \\ = \; & 2N\pi \end{aligned} \tag{8.42}$$

where N is an integer. The resonance frequencies, which depend on the indices N as well as mn, are, from above

$$\nu_{Nmn} = N\frac{c}{2d} + \frac{c}{2\pi d}(m+n+1)\left[\arctan\left(\frac{d_1}{b}\right) + \arctan\left(\frac{d-d_1}{b}\right)\right] \tag{8.43}$$

The first term, ν_{Nmn}, which depends on N only, is the round - trip frequency of a plane wave along the optical axis. Sometimes the index N is called the *axial* or *principal mode number*, and the indices m and n are called the *transverse* or higher-order mode numbers associated with the mode N. A typical spectrum is illustrated in Fig. 8.25. Usually the resonance peaks decrease, and the widths increase, with increasing transverse mode order because of increasing diffracting loss around the mirrors.

A simple explanation for the different resonance frequencies is that the phase shift is a manifestation of the different phase velocities of the different Gaussian beams. In the ray picture, the higher-order modes propagate at larger angles to the axis and the phase velocities are correspondingly larger, leading to higher resonance frequencies.

The resonator is often used as a *scanning Fabry–Perot interferometer* to measure the frequency spectrum of an optical signal. The signal beam is coupled into the resonator, and the length of the resonator d is varied slightly, often less than one wavelength, so that the resonance frequencies are varied slightly. When a resonance frequency coincides with a frequency of the incident beam, the transmission of that frequency from the beam through the resonator is high, as shown in Fig. 8.25. By scanning d, the spectral distribution of the beam can be measured. However, when the beam is coupled into the resonator, unless care is taken to match the beam to one resonator mode, each frequency component

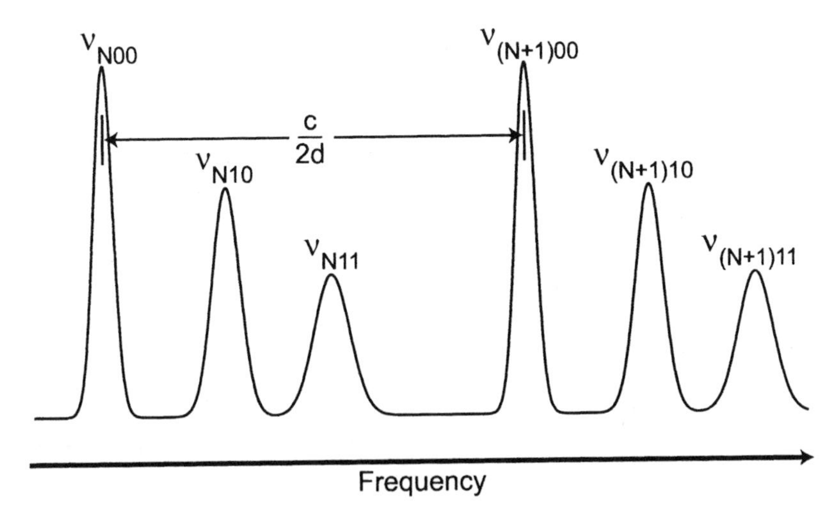

Figure 8.25: Spectrum of a resonator: the spectrum is periodic in frequency. One period is called the *free spectral range.* The principal modes are highest and narrowest. The peaks decrease and the widths increase with transverse mode number.

in the beam will be coupled into many higher-order modes of the resonator, each of which has a different resonance frequency, and measurement by scanning will lead to confusing and erroneous results. This problem can be circumvented by designing the resonator so that all of the higher-order resonances coincide either with an axial mode resonance or with another higher-order resonance. In particular, if the arc tangents in Eq. 8.43 add to $\pi/2$, then the resonance frequencies simplify to

$$\nu_{Nmn} = \frac{c}{4d}\left(2N + m + n + 1\right)$$

The frequencies are periodic, with a periodicity of $c/(4d)$, as if the resonator were doubled in length but with all the higher-order modes banished. If d is adjusted so that the spectral width of the incident beam is smaller than $c/4d$, then the spectrum can be unambiguously measured without careful mode matching into one single mode of the resonator. This design can be realized most simply with a symmetric resonator, $R_1 = R_2$. By symmetry, $d_1 = d/2$ and we require

$$\arctan\left(\frac{d_1}{b}\right) + \arctan\left(\frac{d - d_1}{b}\right) = \frac{\pi}{2}$$

or

$$\arctan\left(\frac{d}{2b}\right) = \frac{\pi}{4}$$

which is to say, $d = 2b$. From Eq. 8.25, this leads to

$$R_1 = R_2 = d$$

This design is called the *confocal resonator.*

Number of modes in a resonator

The transverse dimension of a Hermite–Gaussian beam increases with the mode order. As the mode order increases, eventually the transverse dimension will approach the finite mirror size. When this happens, we take this mode to be the highest-order mode that can be supported by the resonator. This number necessarily varies with each resonator, but we will take the confocal resonator to make an estimate of a typical number.

First, let us estimate the extent to which the value of a Hermite–Gaussian function $H_n(\xi)\exp(-\xi^2/2)$ is significant. Recall that $H_n(\xi)$ is a polynomial of order n, so for large ξ, the dominant term is ξ^n. Approximating the Hermite–Gaussian at large ξ by $\xi^n \exp(-\xi^2/2)$ and setting its derivative to zero, we find the approximate maximum to be at $\xi = \sqrt{n}$. For the Gaussian beam in a confocal resonator, $\xi = \sqrt{2}x/w$. At the mirrors $|z| = d/2 = b$, hence $w = \sqrt{2}w_0$, and the nth-order Hermite–Gaussian mode extends to approximately $\sqrt{n}w_0 = \sqrt{n\lambda b/\pi}$. If the mirror diameter is D and we equate $\sqrt{n\lambda b/\pi}$ to $D/2$, we find

$$n = \pi \left(\frac{D}{2}\right)^2 \frac{1}{\lambda b} = \pi \left(\frac{D}{2}\right)^2 \frac{2}{\lambda d} \sim \frac{\text{mirror area}}{\text{wavelength} \times \text{resonator length}}$$

This number is also called the *Fresnel number*of the resonator.

Frequency selectivity and energy storage

An important property of a resonator is its frequency selectivity. Related to frequency selectivity is energy storage. To illustrate, we consider a particularly simple resonator, the symmetric resonator consisting of two identical mirrors separated by a distance d. The radius of curvature of the mirrors is R, the field reflectivity ρ and field transmittance τ so that, in the absence of diffraction loss around the mirror edge, by energy conservation $|\rho|^2 + |\tau|^2 = 1$. Suppose that an incident Gaussian beam from the left is mode-matched to a resonator mode mn. The Gaussian beam inside the resonator consists of two counterpropagating beams reflected into each other by the mirrors. Let the field of the right-traveling beam be E_1 at the left mirror and E_2 at the right mirror; let the field of the left-traveling beam be E_3 at the right mirror and E_4 at the left mirror. Just outside the left mirror there is the incident beam E_i traveling to the right and the reflected beam E_r traveling to the left, and just outside the right mirror there is the transmitted beam E_t traveling to the right. The fields are illustrated in Fig. 8.26. By symmetry, the beam waist is in the middle of the resonator. The boundary conditions at the left mirror are

$$\begin{aligned} E_1 &= \tau E_i + \rho E_4 \\ E_r &= \tau E_4 \end{aligned}$$

and at the right mirror

$$\begin{aligned} E_3 &= \rho E_2 \\ E_t &= \tau E_2 \end{aligned}$$

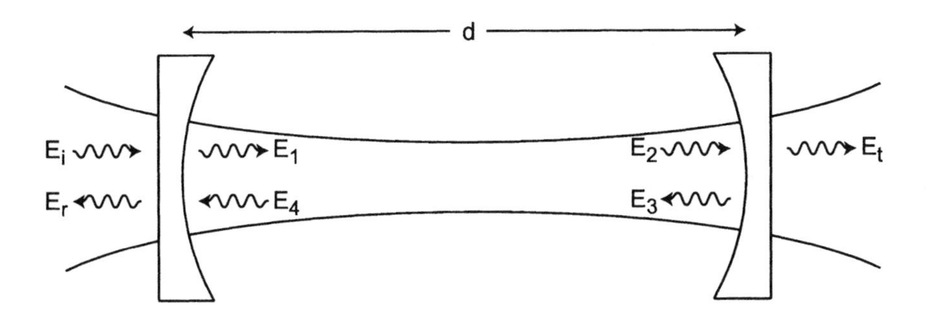

Figure 8.26: Waves of a resonator: shown are the waves at the two mirrors of a resonator, with an incident wave E_i from the left.

Now, because of the symmetry of the resonator,

$$\frac{E_2}{E_1} = \frac{E_4}{E_3} = \exp\left[ikd - 2i\phi_{mn}\left(\frac{d}{2}\right)\right]$$

From these equations, we can find the field inside the resonator, say E_1, in terms of E_i and the resonator parameters:

$$E_1 = \frac{\tau}{1 - \rho^2 \exp\left[2ikd - 4i\phi_{mn}(d/2)\right]} E_i$$

as well as the transmitted field

$$E_t = \frac{\tau^2 \exp\left[ikd - 2i\phi_{mn}\left(d/2\right)\right]}{1 - \rho^2 \exp\left[2ikd - 4i\phi_{mn}\left(d/2\right)\right]} E_i$$

Let $\mathcal{R} = |\rho|^2$, $\mathcal{T} = |\tau|^2$. The ratio of the field intensity inside the resonator to the incident intensity is

$$\frac{|E_1|^2}{|E_i|^2} = \frac{\mathcal{T}}{(1-\mathcal{R})^2 + 4\mathcal{R}\sin^2(kd - \phi_{mn})} \tag{8.44}$$

and the ratio of the transmitted field intensity to the incident intensity is

$$\frac{|E_t|^2}{|E_i|^2} = \frac{\mathcal{T}^2}{(1-\mathcal{R})^2 + 4\mathcal{R}\sin^2(kd - \phi_{mn})} = \mathcal{T}\frac{|E_1|^2}{|E_i|^2} \tag{8.45}$$

The frequency dependence of $|E_1|^2$ and $|E_t|^2$ is in the argument of the sine function, $kd - \phi_{mn}(d/2)$. The frequency dependence of ϕ_{mn} is $\arctan(d/2b)$. For the small fractional frequency changes considered here, the corresponding change in the Rayleigh range b is small. On the other hand, since d is usually many wavelengths, change in kd is substantial. So for the following discussion, all the frequency dependence can be placed on the factor kd. Now suppose that the frequency of the incident beam is such that the argument of the sine in the two expressions Eqs. 8.44 and 8.45 is equal to the resonance condition

$N\pi$, where N is an integer. Further, if the diffraction loss is negligible so that $\mathcal{T} = 1 - \mathcal{R}$, then

$$\frac{|E_1|^2}{|E_i|^2} = \frac{1}{(1-\mathcal{R})} = \frac{1}{\mathcal{T}}$$

and

$$\frac{|E_t|^2}{|E_i|^2} = 1$$

The field inside the resonator builds up to a value much higher than the incident field intensity when the mirror reflectivity is high ($\mathcal{R}$ approaching unity). The intensity enhancement is directly related to the frequency selectivity of the resonator, as seen immediately below. The transmitted field intensity, on the other hand, is equal to the incident field. By energy conservation, the reflected field is zero. It should be noted that total transmission is a result of two necessary conditions: (1) the mirrors have equal reflectivity, and (2) the resonator has no loss other than transmission through the mirrors (absorption or diffraction around the mirror edges is negligible).

The resonator is often used as a frequency-selective element. When the frequency is slightly off resonance, the transmitted field drops off rapidly. When the frequency changes from resonance so that

$$\delta(kd) = \frac{\delta\omega}{c}d = \pm\frac{1-\mathcal{R}}{2\sqrt{\mathcal{R}}}$$

the transmitted intensity drops by half. Since the periodicity of the function sine squared is π, the transmission as expressed in Eq. 8.45 is periodic in angular frequency with a periodicity of $\pi c/d$, or in frequency, $c/(2d)$. This frequency period is called the *free spectral range* (FSR) of the resonator. When the resonator is used to measure the spectrum of a signal, the signal spectrum must be narrower than the FSR, otherwise at any moment the signal is transmitted through at least two resonance peaks and the spectral measurement is invalid. A measure of the frequency selectivity is given by the ratio of the FSR and the full frequency width of the resonance peak. This ratio, called the *finesse*$\mathcal{F}$ of the resonator, is

$$\mathcal{F} \equiv \frac{\pi\sqrt{\mathcal{R}}}{1-\mathcal{R}}$$

The finesse has another simple physical interpretation; it is the number of round trips the entrapped field undergoes before leaving the resonator.[6] Consider the case when R approaches unity. Follow the field intensity as it is reflected between the two mirrors. Let the intensity on the *nth* round trip be $|E|_n^2$. After another round trip, it is

$$|E|_{n+1}^2 = \mathcal{R}^2|E|_n^2$$

Therefore

$$|E|_{n+1}^2 - |E|_n^2 = -(1-\mathcal{R}^2)\,|E|_n^2 = -(1+\mathcal{R})(1-\mathcal{R})\,|E|_n^2 \simeq -2(1-\mathcal{R})|E|_n^2$$

[6] As apparent from a glance at Eq. 8.44, it is also equal, within a factor of π, to the ratio of the intensity inside the resonator on resonance to the incident intensity.

Hence

$$|E|_n^2 \simeq |E|_0^2 \exp\left[-2(1-\mathcal{R})n\right] \simeq |E|_0^2 \exp\left[-2\pi n/\mathcal{F}\right]$$

So that the intensity decays to $e^{-2\pi}$ of its original value after $\mathcal{F}$ round trips.

Related to the finesse is the quality factor Q of the resonator, which is the ratio of the resonance frequency (not the resonance periodicity) to the resonance width, or N times the finesse. But the resonance index N is just the number of half wavelengths in the resonator length d, and therefore Q is the number of oscillations the field intensity undergoes before leaving the resonator.

8.5.2 The multimirror resonator

Many optical resonators have lenses or mirrors between the two end mirrors. The modes and stability of such resonators are determined by the same conditions as for the two-mirror resonator, that is, self-reproduction after one round trip and reality and positiveness of the beam waist. The $ABCD$ matrices facilitate analysis by reducing the problem to a multiplication of matrices. Alternatively, using the fact that the wave curvature must be the same as the mirror radius, one can construct an equivalent two-mirror resonator by successive imaging of all but one of the sequential lenses in the resonator. To illustrate the method, we use one simple example whose solution can be easily obtained directly.

Consider a resonator consisting of two plane end mirrors with a lens in the middle (Fig. 8.27a). Obviously a symmetric system, the beam curvature R at the lens is equal to $\pm 2f$, where f is the focal length of the lens. The beam waist at the plane mirror, or equivalently the Rayleigh range b, is then to be solved from

$$R = d + \frac{b^2}{d}$$

or

$$b = \sqrt{d(R-d)} \tag{8.46}$$

where d is the half length of the resonator.

Now let us solve the same problem by imaging the right-hand end-plane mirror with the internal lens. The image position d_i from the lens is given by the lens law

$$d_i = f\frac{d}{d-f}$$

Recall from our discussion on imaging with a thin lens, that the image acquires a phase curvature of $fd_i/d = f^2/(d-f)$. The image of the plane mirror is therefore a spherical mirror of radius $R' = f^2/(d-f)$, at a distance $d' = d - d_i$ from the left-end mirror. The left end mirror and the image spherical mirror is the equivalent two-mirror resonator of the original three-element resonator (Fig. 8.27b). The Rayleigh range b' of the mode at the plane mirror of the equivalent resonator is given by the same formula as Eq. 8.46

$$b' = \sqrt{d'(R'-d')}$$

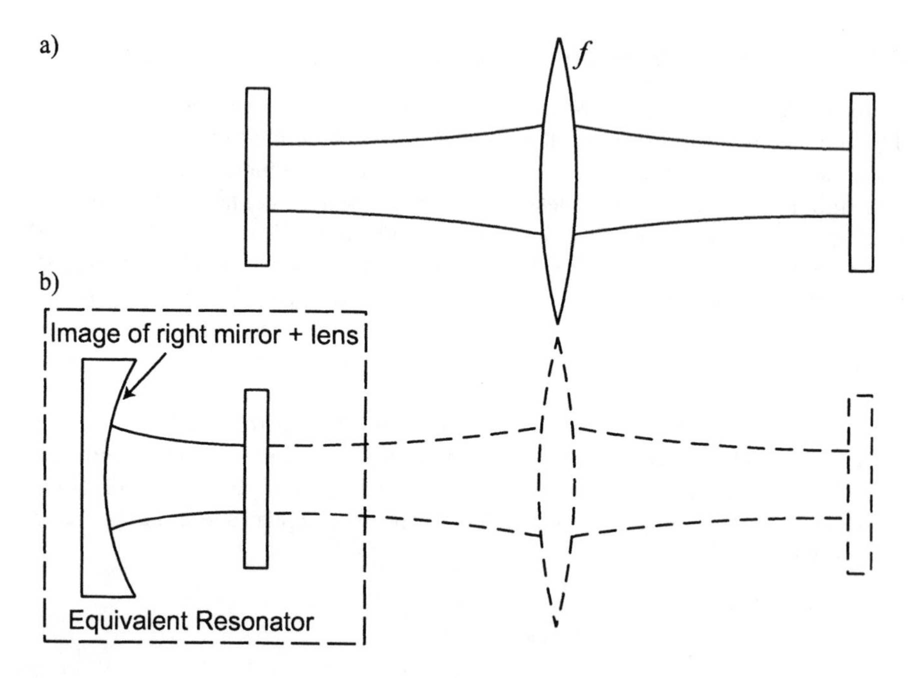

Figure 8.27: A three-element resonator and its equivalent two-mirror resonator: (a) a symmetric resonator with a lens in the middle and plane mirrors at the ends. (b) the equivalent two-mirror resonator. The left, curved mirror is the image of the right-hand plane mirror formed by the lens. The resonators are equivalent in the sense that when the beams inside the resonators are extended to a common region, they are identical.

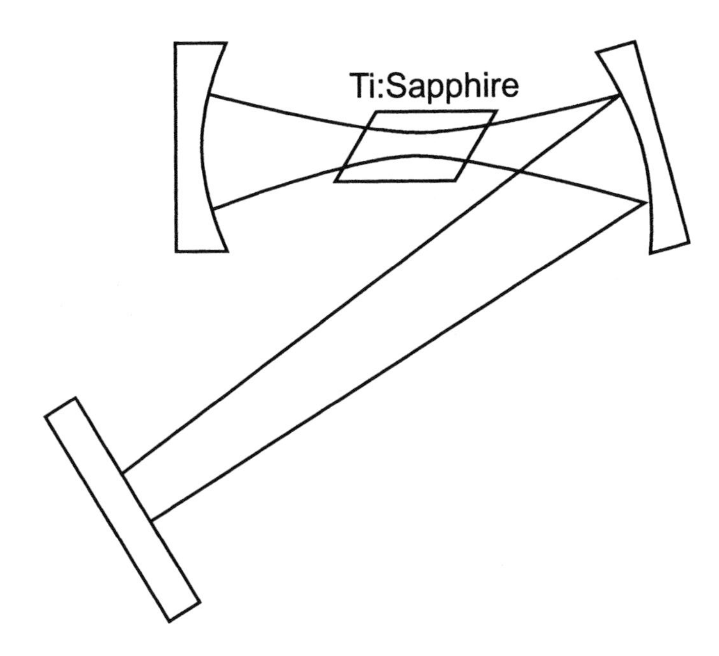

Figure 8.28: Compound resonator design applied to Ti:Sapph laser.

which, after simplification, is

$$b' = \sqrt{d(R-d)}$$

the same as that given by Eq. 8.46. Once the position and magnitude of the beam waist are known, the Gaussian beam is completely characterized.

Problem 8.10 *Laser resonator design:*
Design a resonator for the titanium sapphire laser discussed in Chapter 7. The laser, as illustrated in Fig. 8.28, is to have two curved mirrors and one plane mirror, with the beam waist in the sapphire between the two curved mirrors. The pump beam is to be focused in the sapphire so that it overlaps with the lasing beam as much as possible, with a Rayleigh range designed to be equal to the absorption length. The round trip time in the resonator is to be 10 ns.

8.6 Further Reading

All the topics in this chapter are covered at greater length and depth in

- H. A. Haus, *Waves and Fields in Optoelectronics*, Prentice-Hall, Upper Saddle River, NJ, 1984

A standard text which has educated two generations of laser engineers is

- A. E. Siegman, *Lasers*, University Science Books, Mill Valley, CA, 1986.

The discussion on multi-element resonator follows

- H. Kogelnik, *Imaging of optical mode - resonators with internal lenses*, Bell Sys. Tech. J. **44**, 455-494, March 1965.

References on optical signal diagnostics using intensity correlation functions can be found in

- L. Yan, P.-T. Ho and C. H. Lee, *Ultrashort optical pulses: sources and techniques*, in *Electro-optics Handbook*, 2nd edition, R. W. Waynant and M. N. Ediger, eds., McGraw-Hill, New York, 2000.

Appendixes to Chapter 8

8.A Construction of a Three-Dimensional Beam

A two-dimensional beam propagating in direction z' is, from Eq. 8.9

$$\int d\theta A_x(\theta) \exp\left[ik\theta x + ik\left(1 - \frac{\theta^2}{2}\right)z'\right]$$

where A_x is the angular distribution. This beam has a finite extent in the x-direction but is unconfined in y'. Suppose that we add such two-dimensional waves, each of which has amplitude $A_y(\varphi)d\varphi$ and propagates in a direction z' inclined at an angle φ from z:

$$z' = z\cos(\varphi) + y\sin(\varphi) \simeq z\left(1 - \frac{\varphi^2}{2}\right) + y\varphi$$

The composite wave is then

$$\begin{aligned}
\psi(x, y, z) &= \int d\theta \int d\varphi A_x(\theta) A_y(\varphi) \exp\left[ik\theta x + ik\left(1 - \frac{\theta^2}{2}\right)z'\right] \\
&= \int d\theta \int d\varphi A_x(\theta) A_y(\varphi) \exp\left[ik\theta x + ik\left(1 - \frac{\theta^2}{2}\right)\left(z\left(1 - \frac{\varphi^2}{2}\right) + y\varphi\right)\right] \\
&= e^{ikz} \int d\theta A_x(\theta) \exp\left[ik\theta x - k\frac{\theta^2}{2}\right] \int d\varphi A_x(\varphi) \exp\left[ik\varphi y - ik\frac{\varphi^2}{2}\right]
\end{aligned}$$

where the higher order terms $y\varphi\theta^2/2$ and $\theta^2\varphi^2/4$ in the exponent have been discarded. In this form, x and y are separated. Repeating the derivation that leads to Eq. 8.14, one can prove, from the equation shown above, Eq. 8.18.

8.B Coherence of Light and Correlation Functions

The very high frequencies of light prevent direct electronic measurement. Since the response time of the measurement system is much longer than an optical cycle, some averaging is made when a measurement of light is made. In addition, in many situations, the light itself varies in some random fashion, making its measurement statistical in nature. In quantitative terms, the measurements are expressed as correlation functions of different orders. In practice, correlation functions of first and second orders are sufficient to characterize the light. The correlation functions of the optical field also measure the degree of coherence of light. The more coherent the light, the more predictable it is. Coherence is often classified into temporal and spatial coherence. *Temporal coherence* refers to the degree of predictability of light at a point in space at a later time given its field at one time at the same point in space. *Spatial coherence* refers to the degree of predictability of light at some point in space given the field at another point at the same time. Spatial coherence is further classified into longitudinal and transverse coherence, referred to the direction of propagation. A few examples will follow the definition of the first order correlation function.

The normalized first order correlation function of an optical field $E(\mathbf{r}, t)$ is

$$\frac{\langle E(\mathbf{r},t)E^*(\mathbf{r}',t')\rangle}{\sqrt{|\langle E(\mathbf{r},t)|^2|E^*(\mathbf{r}',t')|^2\rangle}}$$

where the angular brackets denote ensemble average. It turns out that the correlation function depends only on the differences of the arguments, not the individual values, so the correlation function can be written as

$$\frac{\langle E(\mathbf{r},\tau)E^*(0,0)\rangle}{\sqrt{|\langle E(\mathbf{r},t)|^2|E^*(0,0)|^2\rangle}} \equiv F^{(1)}(\mathbf{r},\tau)$$

Two particular cases are familiar. $F^{(1)}(0,0)$ is just the normalized average intensity. The Fourier transform of $F^{(1)}(0,\tau)$ with respect of τ is the power spectrum, a result known as the Wiener–Khinchine theorem, and is the entity measured by the scanning Fabry-Perot discussed in Section 8.5.1. The time difference τ by which $F^{(1)}$ decreases to $\frac{1}{2}$ is called the *correlation time.* The longer the correlation time, the more coherent the field. The inverse of this correlation time, by Fourier transform, is roughly the spectral width of the field. Similarly, the distance over which $F^{(1)}$ decreases to $\frac{1}{2}$ is called the *correlation distance.* Sunlight, viewed from the earth, is from a point source. The light consists of many frequencies, so that in a very short time the field changes completely and the correlation time is extremely short, and the light is usually called *temporally incoherent.* It is spatially coherent in a direction transverse to the line between the sun and the earth, since the difference in propagation time to these two points is shorter than the correlation time; it is spatially incoherent along the line. Temporally incoherent light from an extended source like a lamp is spatially incoherent. If the light is viewed after passing through a very narrowband filter, then it can become temporally coherent as well as spatially coherent over a finite distance. Light from a laser is both temporally and spatially coherent.

The availability of lasers capable of generating ultrashort pulses shows up the inadequacy of $F^{(1)}$ because it cannot distinguish between coherent pulses, noise bursts, and continuous light that fluctuates randomly, if they all have the same power spectrum. Even with coherent pulses, $F^{(1)}$ cannot distinguish those that are transform-limited and those that are not transform-limited. All these cases can be diagnosed with the addition of the second-order correlation function

$$\langle E(0)E^*(0)E(\tau)E^*(\tau)\rangle \equiv \langle I(0)I(\tau)\rangle$$

where I is the intensity and we have omitted the spatial variable as spatial correlation is rarely measured with second order correlation. The intensity correlation functions for coherent pulses, noise bursts, and continuous, random light are shown in Fig. 8.29. Proof is left to the references.

The treatment in this chapter is for single-frequency fields that are temporally coherent. Multifrequency fields can be treated in the same way and the

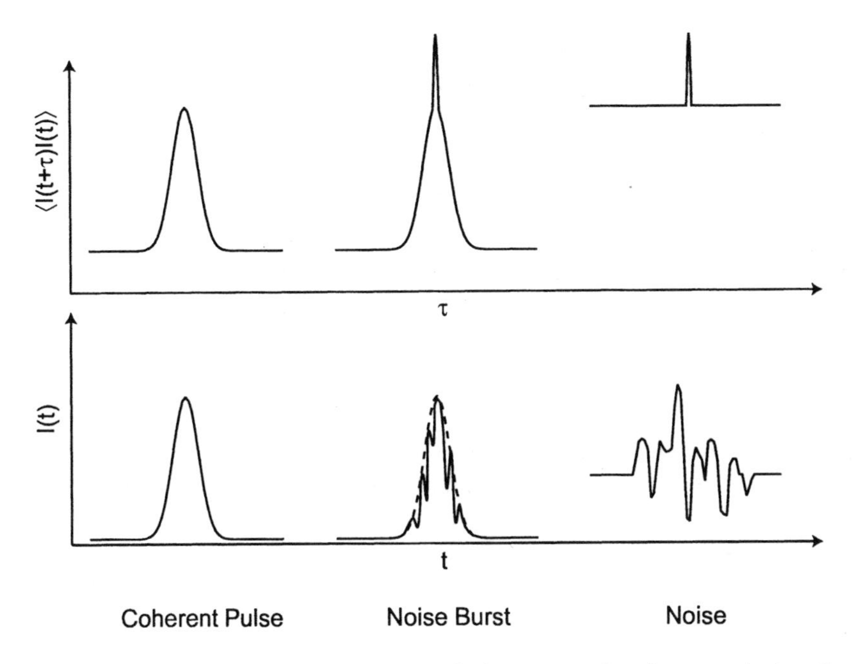

Figure 8.29: Common optical signals and their second-order correlation functions: the upper traces are second-order correlation functions versus time delay. The lower traces are intensities in time.

result obtained by superposition. For temporally incoherent fields, usually the intensity is sought. The cross-frequency terms average to zero in this case.

8.C Evaluation of a Common Integral

An integral used many times in this chapter is

$$I \equiv \int_{-\infty}^{\infty} dx \; \exp\left[-Ax^2 - Bx\right]$$

First, we will evaluate the simpler integral

$$J \equiv \int_{-\infty}^{\infty} dx \; \exp\left[-x^2\right]$$

J can be evaluated by first evaluating its square, J^2, through a transformation from rectangular to polar coordinates:

$$\begin{aligned}
J^2 &= \int_{-\infty}^{\infty} dx \; \exp\left[-x^2\right] \int_{-\infty}^{\infty} dy \; \exp\left[-y^2\right] \\
&= \int_0^{2\pi} d\phi \int_0^{\infty} r dr \exp\left[-r^2\right] = 2\pi \int_0^{\infty} \left(\frac{1}{2}\right) d\left(r^2\right) \exp\left[-r^2\right] \\
&= \pi
\end{aligned}$$

Hence $J = \sqrt{\pi}$.

Now the exponent in the integral of I can be made into a perfect square by adding and subtracting a constant term:

$$Ax^2 + Bx = A\left(x + \frac{B}{2A}\right)^2 - \frac{B^2}{4A}$$

Substituting this into I, changing the integration variable to $x' = \sqrt{A}(x + \frac{B}{2A})$, and using the result $J = \sqrt{\pi}$, one arrives at the final result:

$$\int_{-\infty}^{\infty} dx \, \exp\left[-Ax^2 - Bx\right] = \sqrt{\frac{\pi}{A}} \exp\left[\frac{B^2}{4A}\right]$$

This result also applies when A is a purely imaginary number. In this case, the integrand does not vanish at infinity and the integral is undefined. This difficulty can be circumvented by adding a small, positive real number ϵ to A before integration, then after integration, letting $\epsilon \to 0$.

Index